U.S. Department
Of Transportation

www.nhtsa.gov
National Highway Traffic Safety Administration

DOT HS 809 784

July 2005

Technical Report

Child Passenger Fatalities and Injuries, Based on Restraint Use, Vehicle Type, Seat Position, and Number of Vehicles in the Crash

Published By:
NHTSA's
National Center for Statistics and Analysis
NCSA

NCSA National Center for Statistics and Analysis 400 Seventh St., S.W., Washington, D.C. 20590

1. Report No. DOT HS 809 784	2. Government Accession No.	3. Recipient's Catalog No.
4. Title and Subtitle Child Passenger Fatalities and Injuries, Based on Restraint Use, Vehicle Type, Seat Position, and Number of Vehicles in the Crash		5. Report Date July 2005
		6. Performing Organization Code NPO-121
7. Author(s) Starnes, Marc		8. Performing Organization Report No.
9. Performing Organization Name and Address Mathematical Analysis Division, National Center for Statistics and Analysis National Highway Traffic Safety Administration U.S. Department of Transportation NPO-101, 400 Seventh Street SW. Washington, DC 20590		10. Work Unit No. (TRAIS)
		11. Contract or Grant No.
12. Sponsoring Agency Name and Address Mathematical Analysis Division, National Center for Statistics and Analysis National Highway Traffic Safety Administration U.S. Department of Transportation NPO-101, 400 Seventh Street SW. Washington, DC 20590		13. Type of Report and Period Covered NHTSA Technical Report
		14. Sponsoring Agency Code
15. Supplementary Notes		

16.   Abstract

In both single-vehicle and multi-vehicle crashes, unrestrained children in passenger cars or LTVs (sport utility vehicle, van, or pickup) were more likely to be killed (among fatal crashes) or injured (among non-fatal crashes), as compared to restrained children.

In multi-vehicle fatal crashes, unrestrained children in LTVs were from 2.5 to 5.4 times as likely to be fatally injured as children who were restrained; comparatively, for children in passenger cars, being unrestrained made the child 1.6 to 1.8 times as likely to be fatally injured. Child passenger age categories in this report were 0-3, 4-7, and 8-15 years old.

In fatal crashes, restrained children in passenger cars were more likely to be fatally injured than restrained children in LTVs. Restrained children in the front seat were more likely to be fatally injured than restrained children in second seat. In two-vehicle fatal crashes involving a passenger car and an LTV, a larger percent of passengers traveling in the passenger car were fatally injured than those traveling in the LTV.

The objective of this study is to analyze passenger vehicle crashes involving children up to 15 years old. The Fatality Analysis Reporting System (FARS), and National Automotive Sampling System (NASS) General Estimates System (GES) were consulted to establish restraint usage trends over a 5-year period from 1998 through 2002. The study is intended to provide a better understanding of where to focus future safety efforts designed to improve highway transportation for children.

17. Key Words restraint use, injury severity, seating position, vehicle body type, passenger vehicles, two-vehicle crash	18. Distribution Statement **Document is available to the public through the National Technical Information Service, Springfield, VA 22161 www.ntis.gov**		
19. Security Classif. (of this report) Unclassified	20. Security Classif. (of this page) Unclassified	21. No. of Pages XX	22. Price

Form DOT F 1700.7 (8-72)   Reproduction of completed page authorized

NCSA National Center for Statistics and Analysis   400 Seventh St., S.W., Washington, D.C. 20590

Table of Contents

1.   Executive Summary

This report, published by National Highway Traffic Safety Administration (NHTSA) National Center for Statistics and Analysis (NCSA), provides insight into fatalities and injuries to passenger vehicle occupants, based on a variety of impact attributes. Variables examined include restraint use, injury severity, crash type, vehicle type, seating position, and passenger age. The analysis was based on 1998 through 2002 data from NCSA's Fatality Analysis Reporting System (FARS), a census of fatal motor vehicle crashes; and NCSA's National Automotive Sampling System General Estimates System (GES), which collects data regarding injuries resulting from motor vehicle crashes.

With the exception of Chapter 7 in this report, which examines non-fatal injuries, all charts and tables focus on data collected from crashes where a person was fatally injured. Although an analysis of fatal crashes provides valuable information, patterns seen in these crashes should be taken in context, given that the source of the data is exclusively fatal crashes.

The fatality and injury risks that passengers face vary by restraint use, vehicle type, crash type, and seating position. Passengers are separated into four age categories: 0-3, 4-7, 8-15, and 16 and older. The age category of 16 and older is designed to put the data on child passengers under 16 into perspective, as the report is focused on child passengers.

Crashes are separated into single-vehicle crashes and multi-vehicle crashes. A separate chapter in this report focuses specifically on two-vehicle crashes. Vehicles are categorized into passenger cars, sport utility vehicles, vans, and pickups. The vehicle category of LTV (light trucks and vans) aggregates three of the four vehicle body types: sport utility vehicles, vans, and pickups. Drivers are not included in this report, as the report focuses on passengers located in the front passenger seat (middle and right) or the second seat (left, middle, and right).

Across all crash types examined in this report, being restrained lowered a passenger's chance of being killed and his chance of being injured, compared to when a passenger was unrestrained. In multi-vehicle fatal crashes, unrestrained children in LTVs were from 2.5 to 5.4 times as likely to be fatally injured as children who were restrained; comparatively, for children in passenger cars, being unrestrained made the child 1.6 to 1.8 times as likely to be fatally injured. In fatal crashes, restrained children in passenger cars were more likely to be fatally injured than restrained children in LTVs.

In fatal crashes, restrained children in the front seat were more likely to be fatally injured compared to restrained children in the second seat. In a non-fatal crash, unrestrained passengers are at a higher risk of being injured when they are involved in a single-vehicle crash versus being involved in a multi-vehicle crash.

Among two-vehicle fatal crashes, the types of vehicles involved in the crash play a role in the likelihood that a passenger is fatally injured. In two-vehicle fatal crashes involving a passenger car and an LTV, a larger percent of passengers traveling in the passenger car were fatally injured than those traveling in the LTV.

This report examines these and many other factors pertaining to the safety of children in passenger vehicles. This report does not analyze all variables existing within the databases created by NCSA. Further analyses can be done which expand the understanding of fatality and injury risks facing young children traveling within passenger vehicles. NCSA plans to conduct these analyses and report the findings in the future.

1.1   Conclusions

Conclusions listed below are based on the following categorical specifications included within the FARS and NASS database systems. NCSA will conduct further studies based on these conclusions in order to gain more insight into the fatality and injury risks facing passengers in motor vehicles.

Vehicle Type: Passenger cars, sport utility vehicles, vans, pickups. The vehicle category of LTV (light trucks and vans) aggregates three of the four vehicle body types: sport utility vehicles, vans, and pickups.
Ages: Among fatal crashes: (age 0-3, 4-7, 8-15, 16 and older). Among non-fatal crashes: (age 0-15, 16 and older).
Restraint Use: Restrained (child safety seat, lap and/or shoulder belt), unrestrained
Seating Positions: Front passenger seat (middle and right) and second seat (left, middle, and right). Drivers are not included
Injury Severity: Among fatal crashes (killed, survived). Among non-fatal crashes (injured, not injured).
Crash Type: Single-vehicle crashes and multi-vehicle crashes
Years: 1998 through 2002

<u>Fatal Crashes</u>

➢   In single-vehicle fatal crashes, unrestrained children in either passenger cars or LTVs were between 2 and 3 times as likely to have been fatally injured as compared to restrained children.

➢   In multi-vehicle fatal crashes, unrestrained children in LTVs were from 2.5 to 5.4 times as likely to have been fatally injured as children who were restrained; comparatively, for children in passenger cars, being unrestrained made the child 1.6 to 1.8 times as likely to have been fatally injured.

➢   In single-vehicle fatal crashes, restrained children in passenger cars were roughly 1.5 times as likely to have been fatally injured as restrained children in LTVs.

➢   In multi-vehicle fatal crashes, restrained children in passenger cars were roughly 2.5 times as likely to have been fatally injured as restrained children in LTVs.

➢   In multi-vehicle fatal crashes, unrestrained children in passenger cars were more likely to have been fatally injured than unrestrained children in LTVs; however this difference was much less than the difference seen among restrained children.

> In fatal crashes, restrained children in the front seat were roughly 1.5 times as likely to have been fatally injured compared to restrained children in the second seat.

> In fatal crashes, the relative protection provided by traveling in the second seat (compared to the front seat) was lessened when the passenger is unrestrained.

> In two-vehicle fatal crashes, a much larger percentage of passengers in the struck vehicle were fatally injured (50%) compared to passengers in the striking vehicle (22%).

> In two-vehicle fatal crashes, when an LTV struck a passenger car, 60 percent of the passengers in the struck passenger car were fatally injured, compared to only 9 percent of the passengers in the striking LTV; however, among crashes where a PC struck an LTV, 29 percent of those in the struck LTV were killed, while 45 percent of those in the striking passenger car were killed. Potential reasons for this difference include the fact that LTVs frequently possess a larger mass and higher center of gravity than passenger cars

Non-fatal crashes

> In a non-fatal crash, unrestrained passengers were much more likely to have been injured than restrained passengers, regardless of many other factors involved in the crash (i.e. vehicle type, passenger age, number of vehicles involved in the crash).

> In a non-fatal crash, unrestrained passengers were at a higher risk of being injured when they were involved in a single-vehicle crash versus being involved in a multi-vehicle crash. This may be due to the severity of certain crash types (i.e. rollover, striking a fixed object) that are more likely to occur during single-vehicle crashes.

> In a non-fatal multi-vehicle crash, passengers in an LTV were less likely to be injured than were passengers in a passenger car.

2.   Introduction

From 1998 through 2002, more than 160,000 passenger vehicle occupants were killed and nearly 13 million were injured in traffic crashes. Over two-thirds of these passenger vehicle occupants were drivers, and nearly one-third were passengers.

During this 5-year period, more than 50,000 passengers were killed, and over 4 million passengers were injured. More than 9,000 of these fatalities and 1.3 million of these injuries occurred among passengers up to 15 years old. This report by NCSA examines crashes that led to these child passenger fatalities and injuries. To help put these crashes involving children into perspective, the report also examines crashes involving passengers 16 and older.

The purpose of this report is to:

   o Examine passenger vehicle fatal crashes to determine how a passenger's likelihood of being killed is affected by restraint use, vehicle body type, seating position and the number of vehicles involved in the crash.
   o Examine passenger vehicle non-fatal crashes, to see how these same variables affect the chance of a passenger being injured.
   o Focus on two-vehicle crashes and investigating the differing roles played by the striking vehicle and the struck vehicle.
   o Gain a better understanding of where to focus future safety efforts designed to improve highway transportation for children and adults.

This report analyzes the relationship between a number of factors pertaining to the safety of children in passenger vehicles. The report examines children up through 3 years old, 4 through 7 years old, 8 through 15 years old, and 16 and older. The older age category is intended to put the findings pertaining to children into perspective.

This report looks at several important crash factors. Much perspective can be gained by looking at the association between these crash factors and the role they play in affecting the likelihood that a passenger is killed or injured. For example, does the type of vehicle a child is traveling in play a larger role in the crash when the child is restrained or when the child is unrestrained? Also, is a passenger in a single-vehicle crash more likely to be injured than a passenger in a multi-vehicle crash? Additionally, in two-vehicle fatal crashes, does vehicle type sometimes lead to passengers in the striking vehicle being more likely to be fatally injured than passengers in the struck vehicle? This report will consider these and other issues, adjusting for important crash factors.

3.   Methodology

This report focuses on examining crashes involving children traveling in passenger vehicles.   The child's restraint use and vehicle body type are the primary independent variables, and whether or not the child was killed in the crash is the dependent variable.   The focus of the report is to determine what key factors play a role in the child surviving the crash; therefore, for each crash type, the percentage of restrained and unrestrained children that were fatally injured in the crash was calculated.   While the report is focused on child passengers, passengers of age 16 and older were also examined in order to help put the findings pertaining to children into perspective.

To provide a better perspective on many of the findings regarding fatal crashes, the report also analyzes non-fatal crashes.   The injury severity of passengers in non-fatal crashes is examined.   Passengers in non-fatal crashes are mainly categorized as injured or not injured, while passengers in fatal crashes are categorized as killed or survived.   These variables used to examine fatal crashes are also used to examine non-fatal crashes.   Looking at both fatal and non-fatal crashes helps paint a clearer picture of the role that restraint use, vehicle type, and other variables play in determining the injury severity of children in different crash types.

It was necessary to group crashes according to the number of vehicles involved in the crash.   Crashes were categorized as either single-vehicle crashes or multi-vehicle crashes.   A stratum of multi-vehicle crashes was then examined further by focusing on two-vehicle crashes.   These two-vehicle crashes were analyzed by looking separately at the passengers in the striking vehicle and the passengers in the struck vehicle.   Due to a limited sample size of crashes for each "striking-struck" combination and passenger's age group, the categories for vehicle body type for these two-vehicle crashes were aggregated into two categories: passenger cars, and LTVs (SUVs, vans, and pickups).

Two key data sources were used in this report: the Fatality Analysis Reporting System (FARS) and the National Automotive Sampling System General Estimates System (GES).   FARS collects data on a census of fatal crashes, and GES collects data from police accident reports regarding injuries resulting from police-reported motor vehicle crashes.

Overall, this report looks at these many FARS and GES passenger vehicle crashes in order to teach us more about how certain crash factors affect a passenger's chance of being killed or injured.

Variables

The following provides an overview of the variables used within this report:

- **Vehicle type** – The report examines fatalities and injuries to occupants of passenger vehicles. Passenger vehicles include the following vehicle body types: passenger cars, sport utility vehicles, vans, and pickups. The vehicle body type category of LTV (light trucks and vans) aggregates three vehicle body types: sport utility vehicles, vans, and pickups.

- **Age** – The data are frequently divided into four age categories: 0 through 3 years old, 4 through 7 years old, 8 through 15, and 16 and older. The age category of 16 and older is designed to put the data on child passengers under 16 into perspective, as the report is focused on child passengers. In Chapter 5, the age group 0 through 3 is further separated into less than one year old and 1 through 3. In Chapter 7, the age categories of 0 through 15 and 16 and older are used.

- **Injury Severity** – For all fatal crashes, injury severity is broken down into two categories: Survived and Fatally Injured.

  For non-fatal crashes, GES data are obtained from the police crash reports. The injury severity discussed in Chapter 7 regarding GES crashes is separated into the following four injury severity categories: (A) incapacitating injury, (B) non-incapacitating injury, (C) possible injury, and (O) no injury. These four categories are stratified into two groups: "injured," a combination of categories A, B, and C, and "uninjured," consisting of category O.

  The injury severity variable in the GES database also includes four other categories (fatal injury; died prior to crash; unknown if injured; and injured, severity unknown), which account for a small percentage of the occupants included in GES crashes; these additional categories are not included in this report.

- **Restraint Use** – A passenger is categorized as being restrained if a lap and/or shoulder belt or child restraint was in use at the time of the crash; passengers who were restrained but whose restraint type was unknown are included in this report, and are categorized as restrained. The restraint use of all remaining passengers was categorized as either unrestrained or unknown; passengers whose restraint use was unknown are not included in this report.

- **Seating Position** – Passengers are limited to the following seating positions: front middle seat, front passenger seat, and second seat (left, middle, and right seat). These passenger seat positions were chosen to improve homogeneity between differing vehicles, such as vans and passenger cars; the sample size was only slightly reduced by not including the third and fourth rows.

- **Person Type** – The report is limited to passengers within passenger vehicles in transport. Vehicle drivers, as well as pedestrians, bicyclists, and occupants of motor vehicles not in transport are excluded from this report.

- **Crash Type** – This report examines all fatal crashes, and is not limited only to crashes where a child is fatally injured.

   Crashes were often stratified into single-vehicle (SV) crashes and multi-vehicle (MV) crashes.

   Chapter 8 looks exclusively at two-vehicle crashes, categorized by the vehicle body type of the "striking" vehicle and the "struck" vehicle. Vehicle body type for these two-vehicle crashes is separated into two categories: passenger cars (PC) and light trucks/vans (LTV). Therefore four crash types are examined:
   - a. PC striking PC
   - b. PC striking LTV
   - c. LTV striking PC
   - d. LTV striking LTV

   Rollover crashes are included in this report.

- **Years** – This report examines passenger vehicle fatalities and injuries over the years 1998 through 2002.

Statistical Qualifiers

- While many chapters in this report examine fatal crashes through the use of the FARS database, Chapter 7 examines injury data gathered from non-fatal crashes that is stored in the GES database. The restraint use percentages differ between these two databases, as restraint use is determined by police and may be overreported for survivors.

- As mentioned earlier in the Variables section in Chapter 3, no occupants in the driver's seat position are included in this report. The focus of this report is on passengers 0 through 15 years old, and these passengers very rarely travel in the driver's seat. In order to improve data compatibility between the age categories included in this report, drivers of all ages are not included in this report.

- This report includes fatal motor vehicle crashes where no motor vehicle occupants were fatally injured. An example of this would be the case of a passenger vehicle striking a pedestrian, where the pedestrian was the only person fatally injured in the crash. The distribution of vehicles involved in these types of fatal crashes may not be consistent with the distribution of vehicles in crashes where a vehicle passenger was fatally injured.

- Many crash characteristics that can be examined are not taken into account within this report. The calculations made within this report are not adjusted for all FARS and GES variables, examples of which include driver age, vehicle impact point, and weather at the time of the crash. This report is intended to be a summary of certain key factors surrounding a crash, and is not to be used as a complete examination of passenger vehicle crashes.

- The fatality rates for restrained and unrestrained occupants should not be compared to calculate effectiveness estimates for restraint use. Certain factors involved in the crash were not accounted for, and thus this report cannot be used to estimate the effectiveness of restraint use. For similar reasons, this report should not be used to contradict previous analyses where double-pair comparison methodology was used to control for crash severity.

- Air bags affect the fatality rates of front-seat passengers. This report does not adjust for the presence of air bags in Chapter 6. The comparison of the risks faced by front-seat and second-seat occupants in Chapter 6 is limited to seat position, age of passenger, and restraint use.

4.   **Percent of Passengers Killed and Percent Survived, by Age of Passenger, Restraint Use, Vehicle Body Type, and Number of Vehicles in the Fatal Crash**

In this chapter, fatal crashes from 1998 through 2002 are examined to determine which passengers involved in the crash were fatally injured and which passengers survived. Passengers are separated into four age categories: 0-3, 4-7, 8-15, and 16 and older. The age category of 16 and older is designed to put the data on child passengers under 16 into perspective, as the report is focused on child passengers. Vehicle drivers are not included in this report.

Fatal crashes were stratified into single-vehicle (SV) crashes and multi-vehicle (MV) crashes. Four categories of vehicle body type were examined: passenger cars, SUVs, vans, and pickups. The vehicle body type category of LTV (light trucks and vans) referred to in this chapter aggregates three of the four vehicle body types: SUVs, vans, and pickups.

4.1   Child Passengers, Age 0-3

This section examines child passengers 0-3 years old, who were involved in fatal crashes. Section 4.1.1 focuses on restrained passengers and Section 4.1.2 on unrestrained passengers.

4.1.1   Restrained Passengers, Age 0 - 3

In single-vehicle fatal crashes, 20 percent of restrained children in passenger cars were fatally injured. This percentage was smaller in sport utility vehicles (11%), vans (14%), and pickups (15%). These numbers present a clear pattern: the percentage of restrained child passengers who were fatally injured was higher in a passenger car than in an LTV.

This difference between passenger car and LTV fatalities was much larger in multi-vehicle crashes than in single-vehicle crashes, as shown in Table 1. In multi-vehicle fatal crashes, the percentage of restrained children in LTVs who were fatally injured ranged from 6 percent in pickups up to 10 percent in vans, for children up through 3 years old. This was far less than the 20 percent seen among child occupants of passenger cars.

Table 1. Restrained Child Passengers in Fatal Crashes, Age 0-3, by Survival Status, Vehicle Body Type, and Crash Type

Body Type	Survival Status	Single-Vehicle Crash		Multiple Vehicle Crash	
		Number	Percent (%)	Number	Percent (%)
Passenger Car	Killed	216	20.1	674	20.4
	Survived	857	79.9	2,632	79.6
	Total	1,073	100.0	3,306	100.0
SUV	Killed	54	11.0	62	6.9
	Survived	439	89.0	842	93.1
	Total	493	100.0	904	100.0
Van	Killed	54	14.3	106	10.4
	Survived	323	85.7	915	89.6
	Total	377	100.0	1,021	100.0
Pickup	Killed	27	15.3	37	6.3
	Survived	149	84.7	554	93.7
	Total	176	100.0	591	100.0

Source: National Center for Statistics and Analysis, NHTSA, FARS 1998-2002

Figure 1: Percent Killed, Among Restrained Child Passengers in Fatal Crashes, Age 0-3, by Vehicle Body Type, and Crash Type

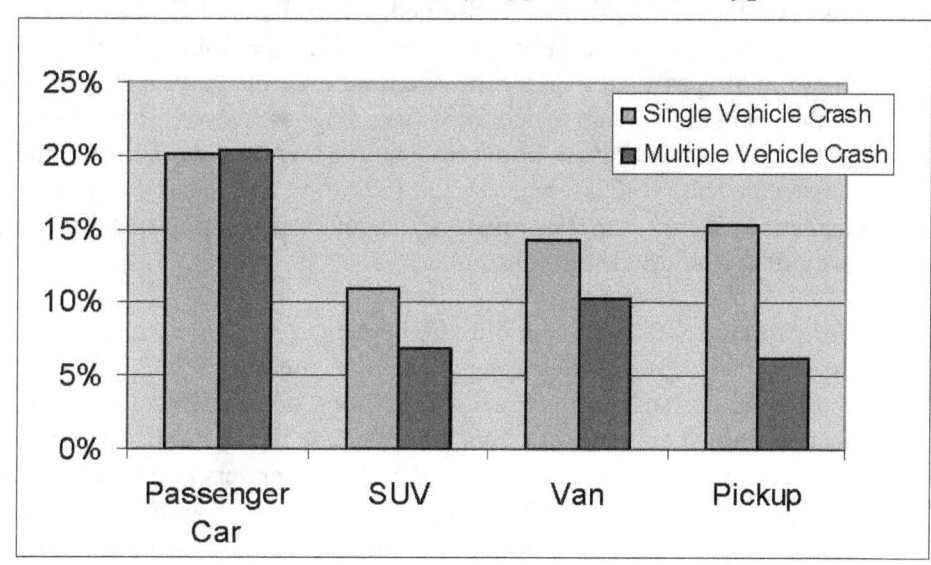

Source: National Center for Statistics and Analysis, NHTSA, FARS 1998-2002

4.1.2   Unrestrained Passengers, Age 0 – 3

For many years NHTSA research has shown that unrestrained passengers involved in fatal crashes are more likely to be fatally injured than passengers who are restrained. Table 2 focuses on unrestrained child passengers and corresponds to the information shown in Table 1 regarding restrained child passengers.

In single-vehicle fatal crashes, the percent of unrestrained children in a passenger car who were fatally injured was more than twice that of restrained children. A comparison of Tables 1 and 2 reflects this percent contrast among children age 0-3 (41% unrestrained versus 20% restrained).

Among unrestrained children age 0-3, Table 2 shows that child passengers in single-vehicle fatal crashes were more likely to be fatally injured in a passenger car (41%) than in a SUV (32%), van (28%), or pickup (34%); as mentioned in Section 4.1.1, restrained children in single-vehicle fatal crashes were also more at risk in passenger cars than LTVs. However, the respective percentages of children killed in a passenger car versus an LTV differed less among unrestrained child passengers than among restrained child passengers.

For a single-vehicle fatal crash, these data suggest that when a child is unrestrained, vehicle body type is less of a determinant of the child's chance of survival than it is when a child is restrained.

In multi-vehicle fatal crashes, the percentage of children fatally injured varied much less across vehicle body type when the child was unrestrained than when the child was restrained. For example, among restrained children 0-3 years old in an MV fatal crash (see Table 1), the percent of children in passenger cars who were killed (20%) was nearly 3 times higher than the percent killed in SUVs (7%). However, Table 2 shows the percent of unrestrained children age 0-3 who were fatally injured in passenger cars (36%) and SUVs (37%) to be nearly equal. Being unrestrained thus removes much of the safety protection that a vehicle can provide a restrained child passenger.

For passengers 0 to 3 years old, the percentage of unrestrained children in an LTV who were killed ranged from 27 percent to 37 percent among multi-vehicle fatal crashes (see Table 2). Comparatively, Table 1 shows that restrained children in LTVs were far less likely to be killed, with only 6 percent to 10 percent of restrained child passengers being fatally injured. The protection that restraint use provides children in multi-vehicle crashes is yet another illustration of the increased safety of being properly restrained.

Table 2. Unrestrained Child Passengers in Fatal Crashes, Age 0-3, by Survival Status, Vehicle Body Type, and Crash Type

Body Type	Survival Status	Single-Vehicle Crash		Multi-Vehicle Crash	
		Number	Percent (%)	Number	Percent (%)
Passenger Car	Killed	181	41.3	278	35.7
	Survived	257	58.7	500	64.3
	Total	**438**	**100.0**	**778**	**100.0**
SUV	Killed	65	32.0	38	36.9
	Survived	138	68.0	65	63.1
	Total	**203**	**100.0**	**103**	**100.0**
Van	Killed	52	27.8	45	27.8
	Survived	135	72.2	117	72.2
	Total	**187**	**100.0**	**162**	**100.0**
Pickup	Killed	60	34.1	47	26.6
	Survived	116	65.9	130	73.4
	Total	**176**	**100.0**	**177**	**100.0**

Source: National Center for Statistics and Analysis, NHTSA, FARS 1998-2002

Figure 2: Percent Killed, Among Unrestrained Child Passengers in Fatal Crashes, Age 0-3, by Vehicle Body Type and Crash Type

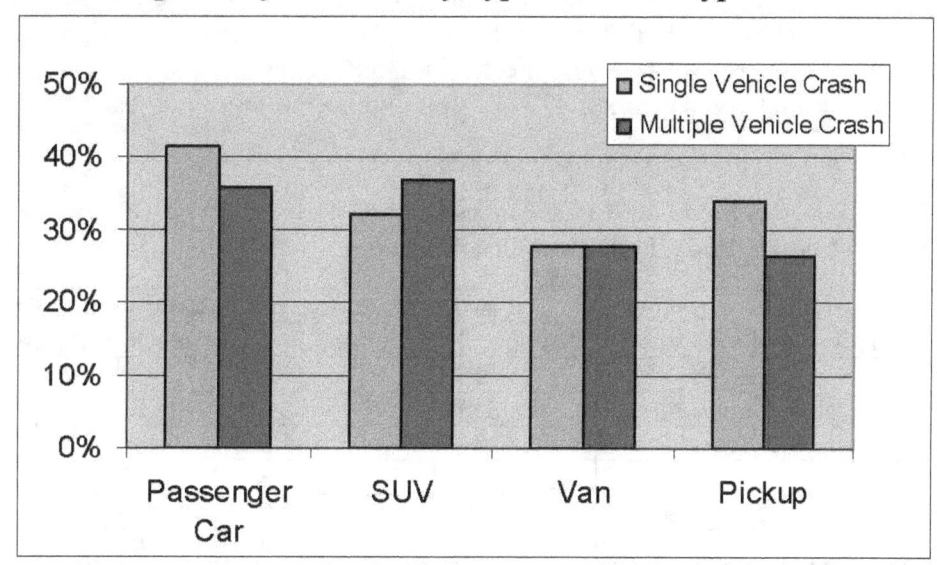

Source: National Center for Statistics and Analysis, NHTSA, FARS 1998-2002

4.2　Child Passengers, Age 4 - 7

This section examines child passengers, 4-7 years old, who were involved in a fatal crash. Section 4.2.1 focuses on restrained passengers and Section 4.2.2 on unrestrained passengers.

4.2.1　Restrained Passengers, Age 4 - 7

In single-vehicle fatal crashes, 14 percent of restrained children in passenger cars were fatally injured. Sport utility vehicles (13%), vans (10%), and pickups (13%) had a lower percentage of children killed. This difference between passenger cars and LTVs was smaller than the difference seen among children in the 0-through-3 age group.

In multi-vehicle fatal crashes, the percent of restrained children in LTVs who were fatally injured ranged from 7 percent in pickups and SUVs up to 8 percent in vans, for children 4 through 7. This was far less than the 18 percent seen among child occupants of passenger cars. As Table 3 displays, the difference between passenger car and LTV fatalities was much larger in multi-vehicle crashes than in single-vehicle crashes.

Among the 190 children 4 through 7 who were killed in single-vehicle crashes, 96 of these fatalities (51%) occurred in passenger cars. Comparatively, out of the total of 595 fatalities that occurred in multi-vehicle fatal crashes, 427 (72%) took place in passenger cars

Table 3. Restrained Child Passengers in Fatal Crashes, Age 4-7, by Survival Status, Vehicle Body Type, and Crash Type

Body Type	Survival Status	Single-Vehicle Crash		Multi-Vehicle Crash	
		Number	Percent (%)	Number	Percent (%)
Passenger Car	Killed	96	13.6	427	17.6
	Survived	612	86.4	2,000	82.4
	Total	**708**	**100.0**	**2,427**	**100.0**
SUV	Killed	42	12.5	52	6.6
	Survived	295	87.5	730	93.4
	Total	**337**	**100.0**	**782**	**100.0**
Van	Killed	29	9.5	76	8.0
	Survived	275	90.5	869	92.0
	Total	**304**	**100.0**	**945**	**100.0**
Pickup	Killed	23	12.6	40	6.8
	Survived	159	87.4	552	93.2
	Total	**182**	**100.0**	**592**	**100.0**

Source: National Center for Statistics and Analysis, NHTSA, FARS 1998-2002

Figure 3: Percent Killed, Among Restrained Child Passengers in Fatal Crashes, Age 4-7, by Vehicle Body Type and Crash Type

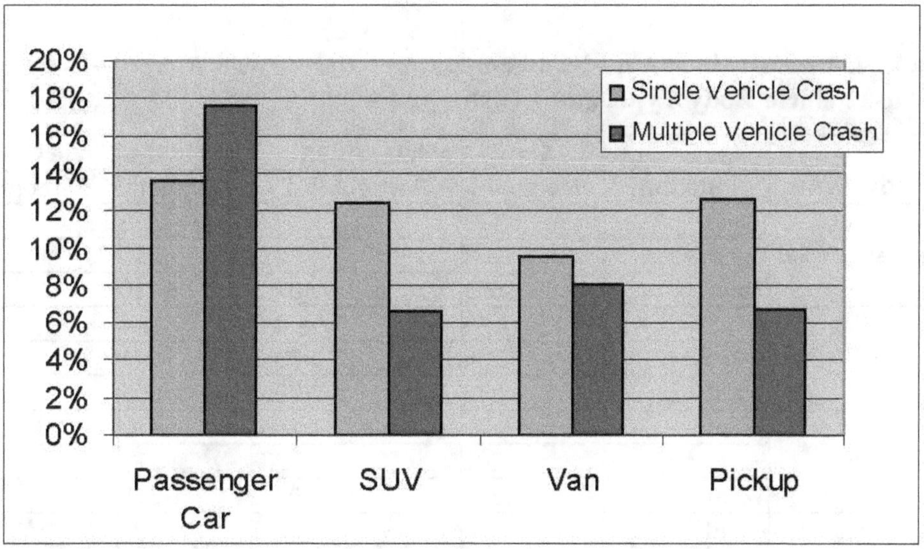

Source: National Center for Statistics and Analysis, NHTSA, FARS 1998-2002

4.2.2   Unrestrained Passengers, Age 4 – 7

A comparison of Tables 3 and 4 shows that the percentage of unrestrained children in a passenger car who were fatally injured (32%) was more than twice that of restrained children (14%), in single-vehicle fatal crashes. This was also true among the 0 through 3 age range, showing that unrestrained passengers involved in a fatal crash are more likely to be fatally injured than passengers who are restrained.

Among unrestrained children 4 through 7, Table 4 suggests that child passengers in single-vehicle fatal crashes were more likely to be fatally injured in a passenger car (32%) than in an SUV (27%), van (28%), or pickup (26%).   As mentioned in Section 4.2.1, restrained children in single-vehicle crashes were also more at risk in passenger cars than LTVs. However, the respective percentages of children killed in a passenger car versus an LTV differed less among unrestrained child passengers than among restrained child passengers. This is more information that suggests that when a child is unrestrained, vehicle body type is less of a determinant of the child's chance of survival than when a child is restrained.

Among restrained children in a MV fatal crash (see Table 3), the percent of children in passenger cars who were killed (18%) was more than twice the percent killed in SUVs (7%), vans (8%), or pickups (7%). Comparatively, the percent of unrestrained children who were fatally injured (see Table 4) in passenger cars (32%) was less than twice the percent seen among LTVs. This pattern among MV fatal crashes is similar to that seen among children 0 through 3 years old.

Table 4. Unrestrained Child Passengers in Fatal Crashes, Age 4-7, by Survival Status, Vehicle Body Type, and Crash Type

Body Type	Survival Status	Single-Vehicle Crash		Multi-Vehicle Crash	
		Number	Percent (%)	Number	Percent (%)
Passenger Car	Killed	178	31.8	328	31.6
	Survived	381	68.2	710	68.4
	Total	**559**	**100.0**	**1,038**	**100.0**
SUV	Killed	93	26.9	54	27.7
	Survived	253	73.1	141	72.3
	Total	**346**	**100.0**	**195**	**100.0**
Van	Killed	79	27.8	70	20.1
	Survived	205	72.2	278	79.9
	Total	**284**	**100.0**	**348**	**100.0**
Pickup	Killed	70	26.2	51	17.5
	Survived	197	73.8	240	82.5
	Total	**267**	**100.0**	**291**	**100.0**

Source: National Center for Statistics and Analysis, NHTSA, FARS 1998-2002

Figure 4: Percent Killed, Among Unrestrained Child Passengers in Fatal Crashes, Age 4-7, by Vehicle Body Type and Crash Type

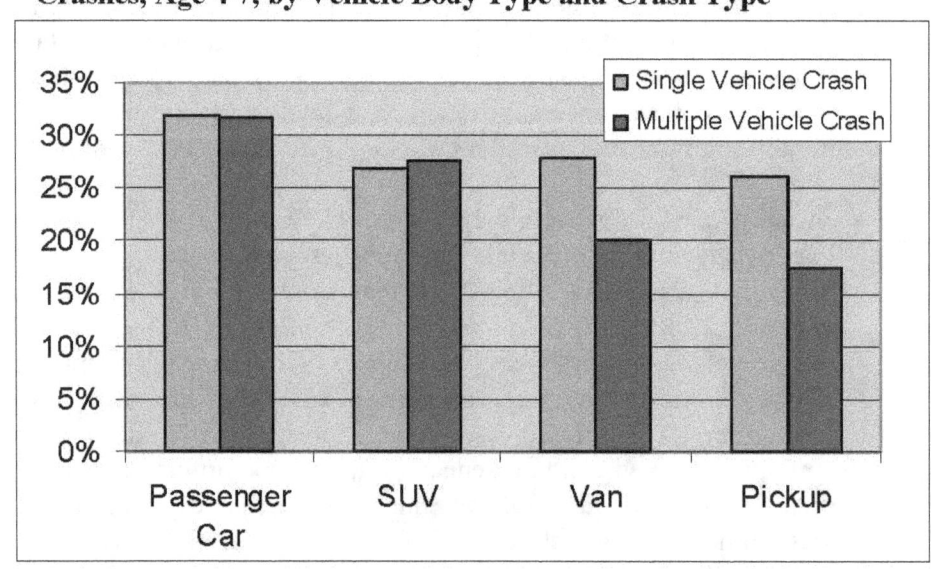

Source: National Center for Statistics and Analysis, NHTSA, FARS 1998-2002

4.3   Child Passengers, Age 8-15

This section examines child passengers 8 through 15 years old who were involved in a fatal crash. Section 4.3.1 focuses on restrained passengers and Section 4.3.2 on unrestrained passengers.

4.3.1   Restrained Passengers, Age 8 - 15

In single-vehicle fatal crashes, 18 percent of restrained children in passenger cars were fatally injured. This percentage was smaller in sport utility vehicles (14%), vans (10%), and pickups (15%). As was true among the two younger age ranges, the percent of restrained child passengers who were fatally injured was higher in a passenger car than in an LTV.

In MV crashes, 18 percent of restrained children in passenger cars were fatally injured. The percent of restrained children in LTVs who were fatally injured ranged from 6 percent in SUVs and pickups up to 7 percent in vans, for children age 8 through 15 years old. This is a repeat of the general trend seen among younger age groups: the difference between the percent of children fatally injured in passenger cars versus LTVs was much larger in multi-vehicle crashes than in single-vehicle crashes, as shown in Table 5.

Out of the 1,150 fatalities among children 8 through 15 years old that occurred in multi-vehicle fatal crashes, 294 (26%) took place in LTVs. Among the 593 children who were killed in single-vehicle crashes, 217 of these fatalities (37%) occurred in LTVs. As seen among other age groups, the percent of fatalities that occurred in LTVs was smaller among MV crashes than SV crashes.

Table 5: Restrained Child Passengers in Fatal Crashes, Age 8-15, by Survival Status, Vehicle Body Type, and Crash Type

Body Type	Survival Status	Single-Vehicle Crash		Multi-Vehicle Crash	
		Number	Percent (%)	Number	Percent (%)
Passenger Car	Killed	376	18.4	856	18.4
	Survived	1,673	81.6	3,801	81.6
	Total	2,049	100.0	4,657	100.0
SUV	Killed	107	13.8	82	5.5
	Survived	669	86.2	1,399	94.5
	Total	776	100.0	1,481	100.0
Van	Killed	56	9.5	136	7.3
	Survived	531	90.5	1,727	92.7
	Total	587	100.0	1,863	100.0
Pickup	Killed	53	11.4	76	5.8
	Survived	412	88.6	1,232	94.2
	Total	465	100.0	1,308	100.0

Source: National Center for Statistics and Analysis, NHTSA, FARS 1998-2002

Figure 5: Percent Killed, Among Restrained Child Passengers in Fatal Crashes, Age 8-15, by Vehicle Body Type and Crash Type

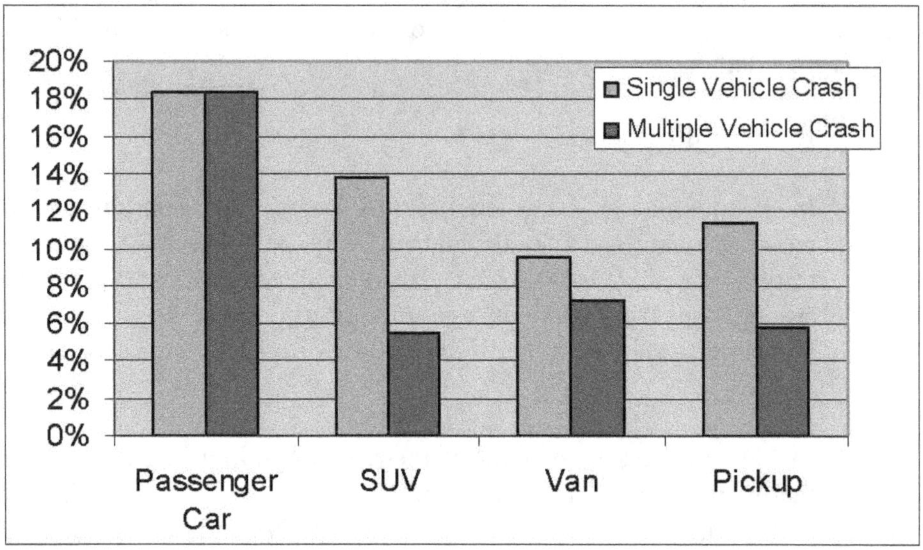

Source: National Center for Statistics and Analysis, NHTSA, FARS 1998-2002

4.3.2   Unrestrained Passengers, Age 8 – 15

In SV fatal crashes, the percent of unrestrained children in a passenger car who were fatally injured was more than twice that of restrained children. Data shown in Tables 5 and 6 reflect this percent contrast among children age 8 through 15 years old (37 percent unrestrained versus 18 percent restrained); also for SUVs (31 percent versus 14%), vans (22 percent versus 10%) and pickups (35 percent versus 11%), the percent of unrestrained children who were fatally injured was more than twice that of restrained children. This same relative risk of above 2.0 was seen among the two younger age categories in this report as well, clarifying once again the large amount of protection provided by restraint use.

The percentage of unrestrained children who were killed ranged from 20 to 23 percent in LTVs, among MV crashes (see Table 6). Comparatively, Table 5 shows that children in LTVs who were restrained were far less likely to be killed, with only 6 percent to 7 percent of restrained child passengers being fatally injured. The protection that restraint use provides children in MV crashes is another indication of the increased safety of being properly restrained.

Table 6. Unrestrained Child Passengers in Fatal Crashes, Age 8-15, by Survival Status, Vehicle Body Type, and Crash Type

Body Type	Survival Status	Single-Vehicle Crash		Multi-Vehicle Crash	
		Number	Percent (%)	Number	Percent (%)
Passenger Car	Killed	999	36.9	850	30.2
	Survived	1,707	63.1	1,961	69.8
	Total	**2,706**	**100.0**	**2,811**	**100.0**
SUV	Killed	373	31.4	150	22.8
	Survived	816	68.6	509	77.2
	Total	**1,189**	**100.0**	**659**	**100.0**
Van	Killed	198	22.4	177	19.8
	Survived	686	77.6	719	80.2
	Total	**884**	**100.0**	**896**	**100.0**
Pickup	Killed	425	35.1	203	20.7
	Survived	786	64.9	778	79.3
	Total	**1,211**	**100.0**	**981**	**100.0**

Source: National Center for Statistics and Analysis, NHTSA, FARS 1998-2002

Figure 6: Percent Killed, Among Unrestrained Child Passengers in Fatal Crashes, Age 8-15, by Vehicle Body Type and Crash Type

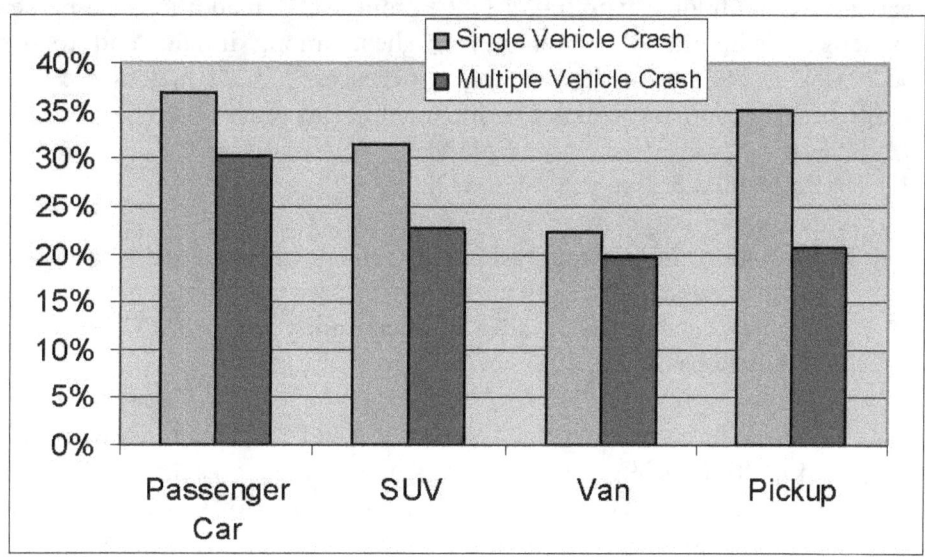

Source: National Center for Statistics and Analysis, NHTSA, FARS 1998-2002

4.4 Vehicle Passengers 16 and Older

This section examines vehicle passengers 16 years and older who were involved in a fatal crash. The drivers of these vehicles were not included in this report. Section 4.4.1 focuses on restrained passengers and Section 4.4.2 on unrestrained passengers.

The data for passengers 16 and older are provided to help place the child occupant data in Sections 4.1 through 4.3 in perspective. Section 4.4 should not be evenly compared with the previous sections. The following example displays this data incompatibility: Among FARS cases included in this report that involved children age 0 through 3, 1,998 children were fatally injured (20%) and 8,177 (80%) survived. For the cases in Section 4.4, 138,815 of these passengers 16 and older were fatally injured (42%) and 190,391 (58%) survived. The percentage of adults (42%) who were fatally injured was more than twice as large as the percentage of children up through 3 (20%) who were fatally injured.

There are many factors in a fatal crash that lead to adult passengers being more likely to be fatally injured than child passengers. For example, passengers seated in the front seat are more likely to be fatally injured than passengers seated in the rear seat, and the percentage of adults traveling in the front seat is higher than the percentage of children who travel in the front seat In addition, the percentage of passengers who travel restrained is highest among infants and toddlers, and decreases as the age of the passenger increases. These factors are beyond the scope of this report and will not be included in this report.

4.4.1 Restrained Passengers 16 and Older

In single-vehicle fatal crashes, 37 percent of restrained passengers 16 and older in passenger cars were fatally injured. As with younger passengers, this percentage was smaller in sport utility vehicles (29%), vans (18%), and pickups (31%).

This difference between passenger car and LTV fatalities was much larger in multi-vehicle crashes than in single-vehicle crashes, as shown in Table 7. In multi-vehicle fatal crashes, the percent of restrained passengers in LTVs who were fatally injured ranged from 15 percent in SUVs and pickups up to 16 percent in vans, which was much lower than the 37 percent seen among occupants of passenger cars.

As was expected due to factors mentioned in the introduction to Section 4.4, the percentage of passengers 16 and older who were fatally injured was much higher than what was seen among children. This was true among all age groups, vehicle body types, and both single-vehicle and multi-vehicle fatal crashes.

Table 7. Restrained Passengers in Fatal Crashes, Age 16 and Older, by Survival Status, Vehicle Body Type, and Crash Type

Body Type	Survival Status	Single-Vehicle Crash		Multi-Vehicle Crash	
		Number	Percent (%)	Number	Percent (%)
Passenger Car	Killed	11,705	36.9	28,768	36.5
	Survived	20,031	63.1	49,974	63.5
	Total	**31,736**	**100.0**	**78,742**	**100.0**
SUV	Killed	2,401	29.0	2,224	14.5
	Survived	5,867	71.0	13,095	85.5
	Total	**8,268**	**100.0**	**15,319**	**100.0**
Van	Killed	928	17.9	2,194	16.1
	Survived	4,244	82.1	11,427	83.9
	Total	**5,172**	**100.0**	**13,621**	**100.0**
Pickup	Killed	2,716	30.6	3,764	14.9
	Survived	6,156	69.4	21,545	85.1
	Total	**8,872**	**100.0**	**25,309**	**100.0**

Source: National Center for Statistics and Analysis, NHTSA, FARS 1998-2002

Figure 7: Percent Killed, Among Restrained Passengers in Fatal Crashes, Age 16 and Older, by Vehicle Body Type and Crash Type

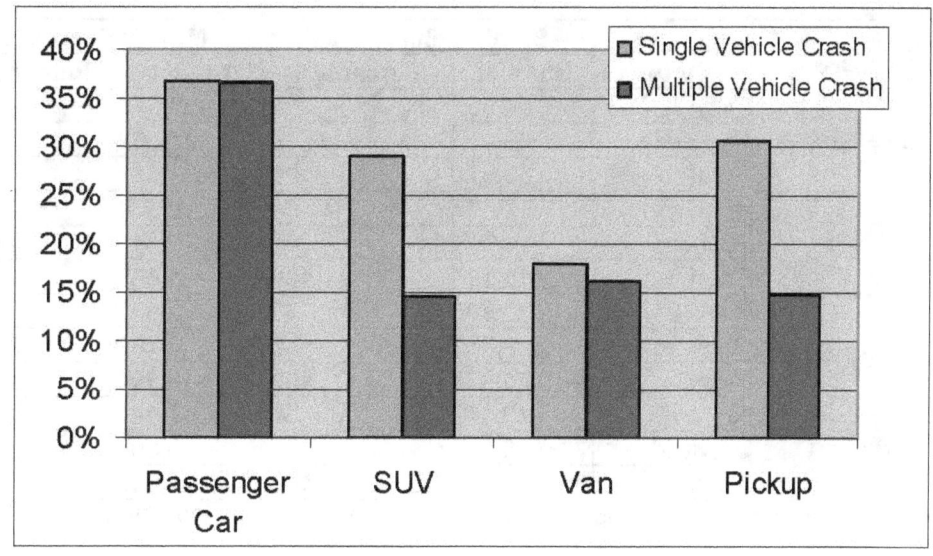

Source: National Center for Statistics and Analysis, NHTSA, FARS 1998-2002

4.4.2 Unrestrained Passengers 16 and Older

Table 8 focuses on unrestrained passengers 16 and older, and corresponds to the information shown in Table 7 regarding restrained passengers. In single-vehicle fatal crashes, the percent of unrestrained passengers in a passenger car who were fatally injured (67%) was 30 percentage points more than that of restrained passengers (37%). Among unrestrained passengers 16 and older, occupants in single-vehicle crashes were more likely to be fatally injured in a pickup (69%) than in a passenger car (67%), SUV (64%), or van (50%).

For passengers 16 and older, the percentage of unrestrained passengers who were killed ranged from 38 percent in vans up to 59 percent in passenger cars, among multi-vehicle crashes (see Table 8). Comparatively, Table 7 shows that restrained passengers were far less likely to be killed in MV crashes, with only 28 percent overall being fatally injured.

Table 8. Unrestrained Passengers in Fatal Crashes, Age 16 and Older, by Survival Status, Vehicle Body Type, and Crash Type

Body Type	Survival Status	Single-Vehicle Crash		Multi-Vehicle Crash	
		Number	Percent (%)	Number	Percent (%)
Passenger Car	Killed	26,396	66.7	22,075	58.7
	Survived	13,175	33.3	15,502	41.3
	Total	**39,571**	**100.0**	**37,577**	**100.0**
SUV	Killed	7,157	64.1	2,552	44.8
	Survived	4,004	35.9	3,143	55.2
	Total	**11,161**	**100.0**	**5,695**	**100.0**
Van	Killed	2,885	49.5	2,322	38.0
	Survived	2,939	50.5	3,782	62.0
	Total	**5,824**	**100.0**	**6,104**	**100.0**
Pickup	Killed	13,440	68.5	7,182	44.8
	Survived	6,168	31.5	8,858	55.2
	Total	**19,608**	**100.0**	**16,040**	**100.0**

Source: National Center for Statistics and Analysis, NHTSA, FARS 1998-2002

Figure 8: Percent Killed, Among Unrestrained Passengers in Fatal Crashes, 16 and Older, by Vehicle Body Type and Crash Type

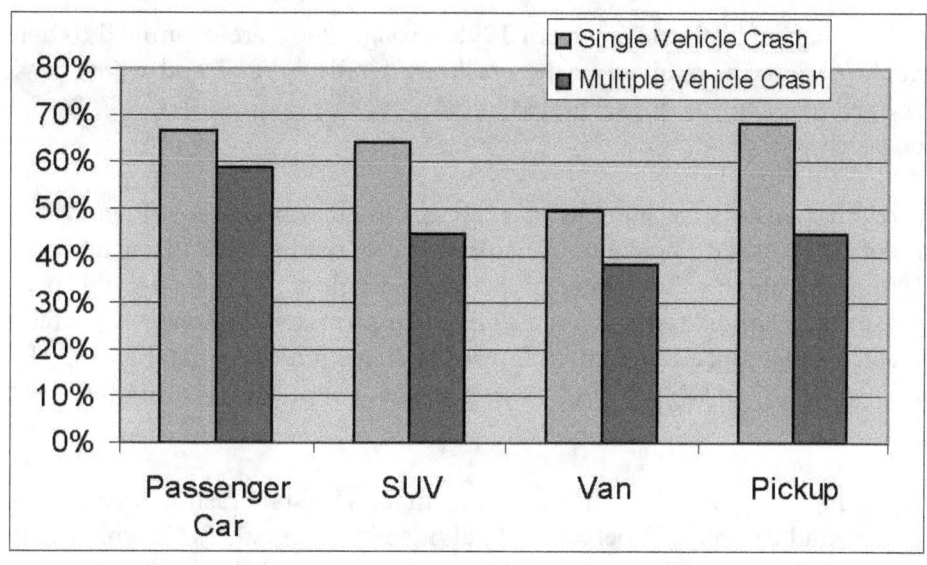

Source: National Center for Statistics and Analysis, NHTSA, FARS 1998-2002

5.   **Relative Risk of Unrestrained Passengers in a Fatal Crash Being Fatally Injured, Compared to Restrained Passengers**

In this chapter, fatal crashes from 1998 through 2002 are examined to determine which passengers involved in the crash are fatally injured and which passengers survived. Passengers are separated into four age categories: 0-3, 4-7, 8-15, and 16 and older.

A relative risk is defined as a ratio of two probabilities, P_1 and P_2. The probabilities used to calculate the relative risks below were discussed in Chapter 4. For calculating the relative risks shown below, P_1 refers to the percent of unrestrained children who were fatally injured in a fatal crash. P_2 refers to the percent of restrained children who were fatally injured in a fatal crash. For each relative risk, P_1 and P_2 pertain to the same age group, vehicle body type, and crash type.

For example, P_1 for 4 though 7 year olds in an SV fatal crash in an SUV was 26.9 percent, and P_2 was 12.5 percent. P_1 divided by P_2 equals 2.15, which shows that for the specific type of crash described, an unrestrained child was 2.15 times more likely to be fatally injured than a restrained child.

For the each of the relative risks shown below, a baseline relative risk of 1.0 is listed in the same row. This is a reminder that each relative risk (i.e. 2.15) was a comparison of the probability of an unrestrained child being fatally injured to the probability of a restrained child being fatally injured, after controlling for age of the child, vehicle body type, and crash type (SV versus MV).

These relative risks for single-vehicle crashes are shown below in Table 9, and the relative risks for multi-vehicle crashes are shown in Table 10.

Relative Risk of Being Fatally Injured While Unrestrained

Data illustrating the benefits of restraint use are mentioned repeatedly throughout this report. These benefits are more consistent among the different vehicle body types for single-vehicle crashes than they are among multi-vehicle crashes.

For nearly all vehicle body types, the percent of unrestrained children in single-vehicle fatal crashes who were killed was consistently between 2 and 3 times as high as the percent of restrained children who were killed (see Table 9). The lowest relative risk was 1.81, representing passengers 16 and older in passenger cars, and the highest relative risk was 3.08, seem among children 8 to 15 years old traveling in vans. Among these SV fatal crashes, 13 of the 16 relative risks were between 2 and 3.

Table 9: Relative Risk of Unrestrained Passengers in a Single-Vehicle Fatal Crash Being Fatally Injured, By Vehicle Body Type and Age of Passenger

Restrained Passengers as Baseline for Each Vehicle Body Type and Age Group

Body Type	Age of Passenger	Unrestrained	Restrained
Passenger Car	0 to 3	2.05	1.0
	4 to 7	2.34	1.0
	8 to 15	2.01	1.0
	16+	1.81	1.0
SUV	0 to 3	2.91	1.0
	4 to 7	2.15	1.0
	8 to 15	2.28	1.0
	16+	2.21	1.0
Van	0 to 3	1.94	1.0
	4 to 7	2.93	1.0
	8 to 15	2.36	1.0
	16+	2.77	1.0
Pickup	0 to 3	2.23	1.0
	4 to 7	2.08	1.0
	8 to 15	3.08	1.0
	16+	2.24	1.0

Source: National Center for Statistics and Analysis, NHTSA, FARS 1998-2002

Figure 9: Relative Risk of Unrestrained Passengers in a Single-Vehicle Fatal Crash Being Fatally Injured, By Vehicle Body Type and Age of Passenger

Restrained Passengers as Baseline for Each Vehicle Body Type and Age Group

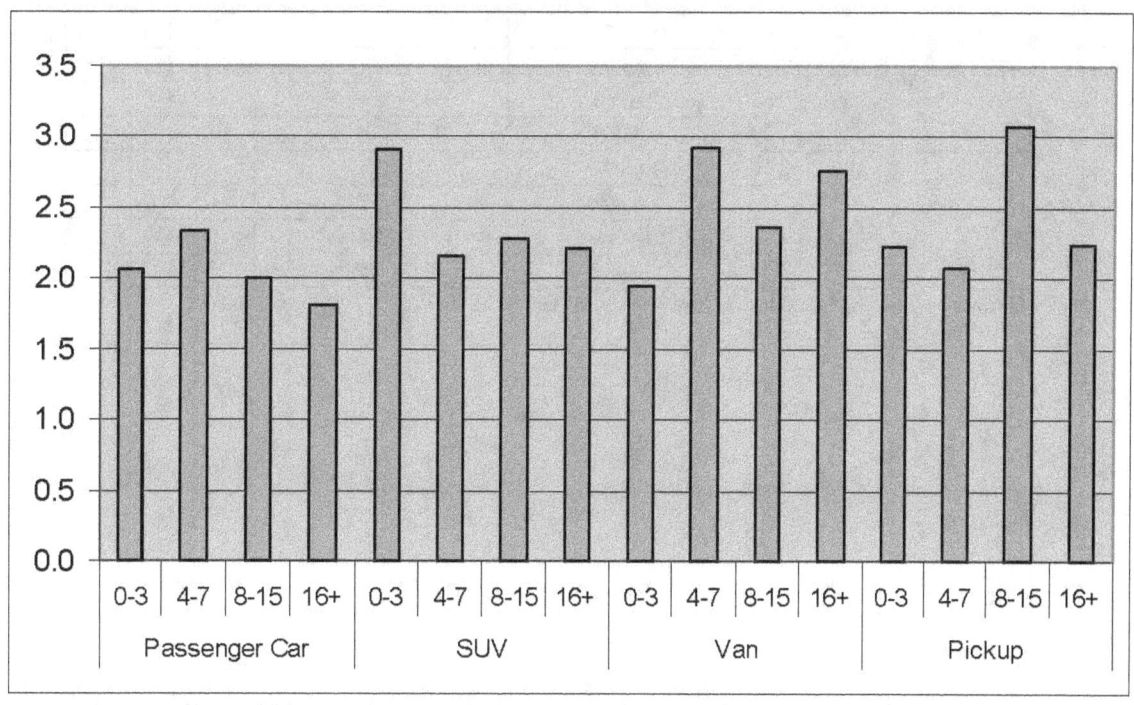

Source: National Center for Statistics and Analysis, NHTSA, FARS 1998-2002

In comparison to the above table and chart regarding single-vehicle crashes, the relative risks for multi-vehicle crashes differed more significantly by vehicle body type. Table 10 shows the relative risks for unrestrained passengers versus restrained passengers in multi-vehicle fatal crashes. For passenger cars, these relative risks fell in the narrow range of 1.61 (16 and older) to 1.80 (4-7-year-olds). This was far less than the relative risks among sport utility vehicles, vans, or pickups, which ranged from a low of 2.36 (16 and older in vans) up to a high of 5.35 (0-3-year-olds in sport utility vehicles). SUVs had the highest relative risks for each of the four age categories. These relative risk calculations on multi-vehicle crashes suggest that being unrestrained may remove much of the protection that an LTV provides to a restrained child passenger.

Table 10: Relative Risk of Unrestrained Passengers in a Multi-Vehicle Fatal Crash Being Fatally Injured, By Vehicle Body Type and Age of Passenger

Restrained Passengers as Baseline for Each Vehicle Body Type and Age Group

Body Type	Age of Passenger	Unrestrained	Restrained
Passenger Car	0 to 3	1.75	1.0
	4 to 7	1.80	1.0
	8 to 15	1.64	1.0
	16+	1.61	1.0
SUV	0 to 3	5.35	1.0
	4 to 7	4.20	1.0
	8 to 15	4.15	1.0
	16+	3.09	1.0
Van	0 to 3	2.67	1.0
	4 to 7	2.51	1.0
	8 to 15	2.71	1.0
	16+	2.36	1.0
Pickup	0 to 3	4.22	1.0
	4 to 7	2.57	1.0
	8 to 15	3.57	1.0
	16+	3.01	1.0

Source: National Center for Statistics and Analysis, NHTSA, FARS 1998-2002

Figure 10: Relative Risk of Unrestrained Passengers in a Multi-Vehicle Fatal Crash Being Fatally Injured, By Vehicle Body Type and Age of Passenger

Restrained Passengers as Baseline for Each Vehicle Body Type and Age Group

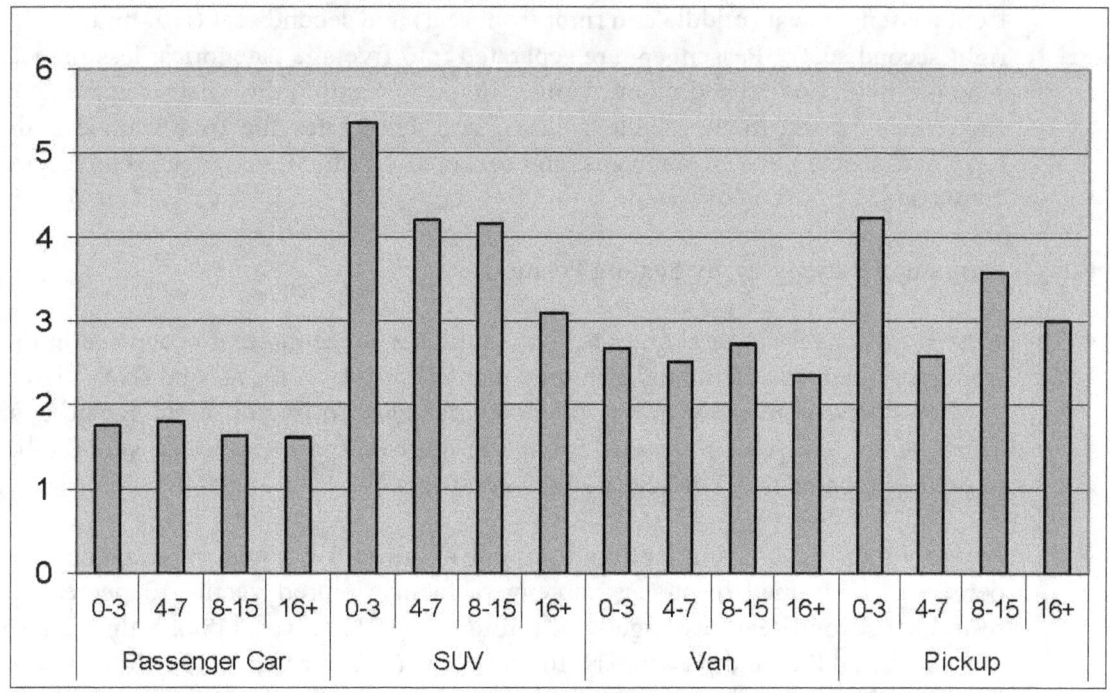

Source: National Center for Statistics and Analysis, NHTSA, FARS 1998-2002

6.   **Percent of Passengers Killed, by Seating Position, Age of Passenger, and Restraint Use, in Fatal Crashes**

The seating position mentioned in this chapter is separated into two categories: front passenger seat (middle and right front seat) and second seat (left, middle and right second seat). Passengers are separated into five age categories: less than 1, 1-3, 4-7, 8-15, and 16 and older. Unlike chapters 4 and 5, this chapter aggregates single-vehicle and multi-vehicle crashes, and aggregates the four vehicle body types (passenger cars, SUVs, vans, and pickups) of which passenger vehicles are comprised.

6.1   Restrained Passengers, by Seating Position

Among passengers in fatal crashes, a much higher percentage of occupants in the front passenger seat were fatally injured than occupants in the second seat. Figure 11 shows these percentages for restrained passengers in five different age groups. Thirty-three percent of restrained infants in the front passenger seat were fatally injured, compared to 20 percent in the second seat.

Figure 11 displays a similar trend among children 1 through 3 years old (18 percent of restrained front-seat passengers fatally injured versus 13 percent of restrained second-seat passengers), 4 through 7 (19% versus 11%), 8 through 15 (16% versus 11%) and passengers 16 and older (27% versus 17%). In each age bracket, a larger percentage of restrained passengers in the front seat were fatally injured compared to restrained passengers traveling in the second seat.

Figure 11: Percent Killed, Among Restrained Passengers in Fatal Crashes, By Age and Seat Position

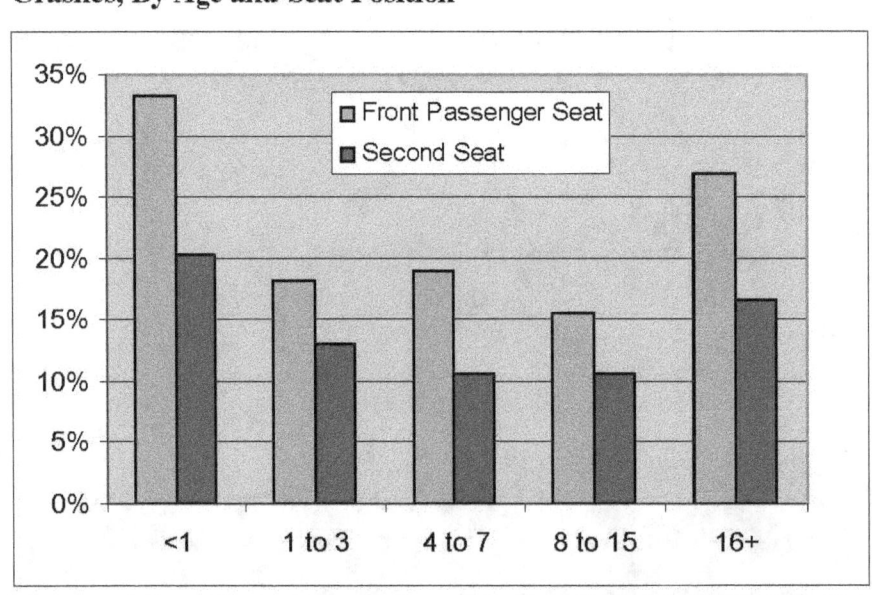

Source: National Center for Statistics and Analysis, NHTSA, FARS 1998-2002

6.2   Unrestrained Passengers, by Seating Position

Figure 12 shows the percentage of unrestrained passengers who were fatally injured, parallel to the information on restrained passengers displayed in Figure 11. Two predictable patterns are quickly evident; first, a much larger percentage of passengers, in both seating positions, are fatally injured when they are unrestrained; and secondly, being unrestrained reduces some of the relative protection that is provided by the second seat compared to the front passenger seat.

A comparison of Figures 11 and 12 shows that restrained infants are 39 percent less likely to be killed in the second seat than in the front seat (20 percent versus 33%), while unrestrained infants are only 13 percent less likely (54 percent versus 62%). For passengers 16 and older, those restrained are 39 percent less likely to be killed when traveling in the second seat compared to the front seat (17 versus 27%), while unrestrained passengers are only 29 percent less likely (35 versus 49%). These two age groups display the pattern that the relative protection provided by traveling in the second seat (compared to the front passenger seat) is lessened when the passenger is unrestrained.

Figure 12: Percent Killed, Among Unrestrained Passengers in Fatal Crashes, By Age and Seat Position

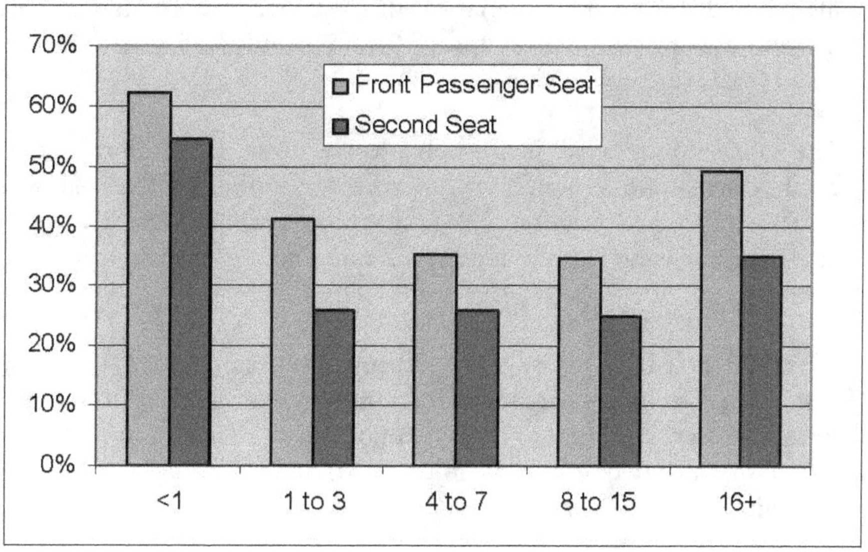

Source: National Center for Statistics and Analysis, NHTSA, FARS 1998-2002

7.   **Percent of Passengers Injured, by Age of Passenger, Restraint Use, Vehicle Body Type, and Number of Vehicles in the Crash**

In this chapter, passenger injury data from 1998 through 2002 are analyzed from the National Automotive Sampling System General Estimates System (GES) database to determine the percent of passengers injured in different types of crashes. These percentages are stratified according to passenger age, passenger restraint use, vehicle body type, and the number of vehicles in the crash.

The GES cases examined in this chapter include a raw total of 169,490 passengers. These cases are weighted to represent the 4.3 million passengers from 1998 through 2002 who were injured in traffic crashes, as well as the 15.6 million passengers who were not injured.

The tables and charts in this chapter focus on the means and standard errors of the percentage of passengers that were injured in various types of crashes. Passengers are separated into two age categories: 0 through 15, and 16 and older. Vehicle drivers are not included in this report.

The passenger age groups of 0-3, 4-7, and 8-15 that have been used throughout this report were aggregated into the age group of 0 through 15 years old in order to create a larger sample size, and therefore a smaller standard error, for each of the many differing crash scenarios; this was because the standard errors that were measured using the age groups of 0-3, 4-7 and 8-15 were too large to produce results that were conclusive.

The crashes examined in this chapter were stratified into single-vehicle (SV) crashes and multi-vehicle (MV) crashes. Section 7.1 focuses on single-vehicle (SV) crashes and Section 7.2 focuses on multi-vehicle (MV) crashes. Four categories of vehicle body type were examined: passenger cars, SUVs, vans, and pickups.

The GES non-fatal injury data presented in this chapter was defined using the following four injury categories: (A) incapacitating injury, (B) non-incapacitating injury, (C) possible injury, and (O) no injury. These four categories have been stratified into two groups: "Injured," made of categories A, B, and C, and "Uninjured," made of category O.

The standard errors presented in this chapter were calculated using SUDAAN, which is a single program consisting of a family of procedures used to analyze data from complex surveys and other observational and experimental studies involving cluster-correlated data.

The Role of Restraint Use in Non-Fatal Crashes Compared to Fatal Crashes

When examining the findings on non-fatal crashes presented in this chapter, it is helpful to consider the role of restraint use in fatal crashes compared to non-fatal crashes. The types of crashes that comprise the group of fatal crashes affect the relative risk of an unrestrained passenger being killed in a fatal crash, compared to a restrained passenger. Some of the fatal crashes are unsurvivable, even if the passengers are properly restrained, with certain high-delta-V crashes being an example of this type of crash. Therefore, in some severe fatal crashes, proper restraint use can sometimes fail to prevent a passenger from being killed.

Comparatively, non-fatal crashes are often moderate enough that they are more likely to allow restraint use to greatly reduce the chance of a passenger being injured. Restraint use plays a large role in these less severe, non-fatal crashes, as it prevents many passengers from being injured.

The above crash factors should be kept in mind when examining (1) the relative risk of an unrestrained passenger being killed in a fatal crash, compared to a restrained passenger, as shown in Chapter 5, and (2) comparing the percentage of unrestrained passengers who were injured with the percentage of restrained passengers who were injured, as is presented in this chapter.

7.1   Passengers in Single-Vehicle Crashes

This section examines passengers who were involved in single-vehicle non-fatal crashes. Section 7.1.1 focuses on passengers age 0 through 15 and Section 7.1.2 on passengers 16 and older.

7.1.1   Passengers in Single-Vehicle Crashes, Age 0 - 15

For passengers 0 through 15 years old in single-vehicle crashes, the percent of restrained passengers who were injured was significantly less than the percent of unrestrained passengers who were injured. This result is seen for all four vehicle body types shown in Figure 13.

Among restrained passengers, the percent who were injured ranged from 17 percent in vans, up to 20 percent in pickups and passenger cars; comparatively for unrestrained passengers, 43 percent of children in vans were injured compared to 52 percent in passenger cars.

For all four vehicle body types, children 0 through 15 were from 2.4 to 2.6 times as likely to be injured when they were unrestrained versus when they were restrained. These injury findings are another example of the safety benefits that are provided by restraint use.

Table 11: Percent and Standard Error of Passengers Injured, Age 0 - 15, By Vehicle Body Type and Restraint Use, in Single-Vehicle Crashes

Body Type	Restrained		Unrestrained	
	Mean (%)	SE (%)	Mean (%)	SE (%)
Passenger Car	19.5	1.4	51.6	2.9
SUV	19.2	2.1	46.8	4.5
Van	16.6	3.7	42.7	7.8
Pickup	19.7	2.9	50.3	3.8

Source: National Center for Statistics and Analysis, NHTSA, GES 1998-2002
Note: These estimates do not include fatally injured passengers.

Figure 13: Percent of Passengers Injured, Age 0 – 15, by Vehicle Body Type And Restraint Use, in Single-Vehicle Crashes

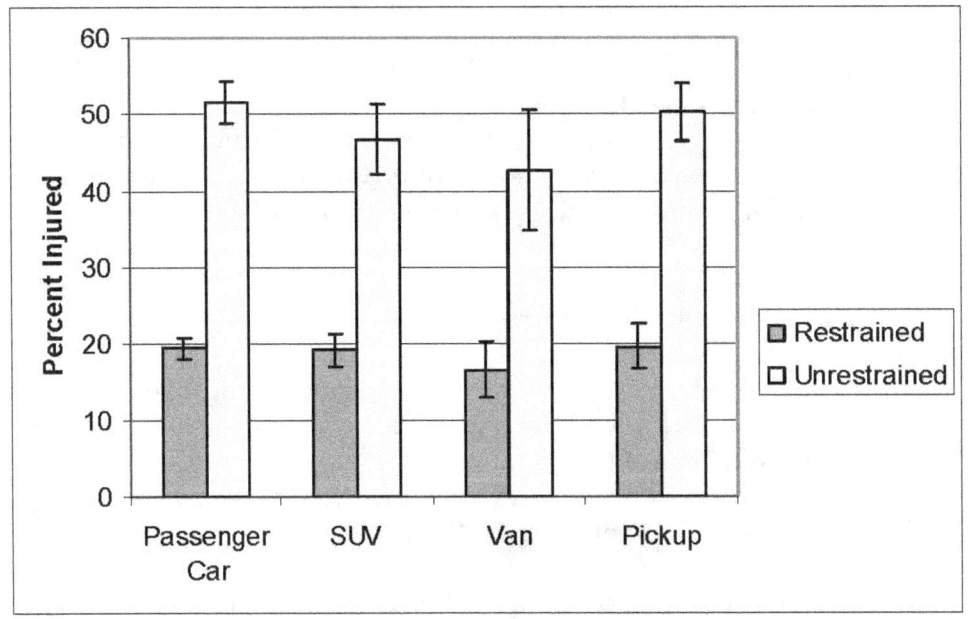

Source: National Center for Statistics and Analysis, NHTSA, GES 1998-2002
Note: Intervals surrounding percent estimates are +/- one standard error. These estimates do not include fatally injured passengers.

7.1.2 Passengers 16 and Older in Single-Vehicle Crashes

For passengers 16 and older in single-vehicle crashes, restrained passengers were much less likely to be injured than were unrestrained passengers. This result, shown below in Figure 14, is similar to the findings on younger passengers shown in Figure 13.

For both restrained and unrestrained passengers 16 and older, the passengers traveling in vans were the least likely to be injured. This same result was seen in Section 7.1.1 for the younger age group. Figure 14 shows that passengers 16 and older in SUVs, among both restrained and unrestrained passengers, were the most likely to be injured. The percent of passengers 16 and older who were injured ranged from 19 percent of those restrained in vans up to 65 percent of those unrestrained in SUVs.

For all four vehicle body types, passengers 16 and older were 2.3 to 2.6 times as likely to be injured when they were unrestrained versus when they were restrained; these relative risks are nearly equal to those seen among passengers under 16 in single-vehicle crashes.

Table 12: Percent and Standard Error of Passengers Injured, 16 and Older, By Vehicle Body Type and Restraint Use, in Single-Vehicle Crashes

Body Type	Restrained		Unrestrained	
	Mean (%)	SE (%)	Mean (%)	SE (%)
Passenger Car	24.5	1.9	55.0	2.3
SUV	26.1	1.9	65.2	2.2
Van	18.8	3.1	43.0	5.5
Pickup	22.3	1.8	58.2	3.2

Source: National Center for Statistics and Analysis, NHTSA, GES 1998-2002
Note: These estimates do not include fatally injured passengers.

Figure 14: Percent of Passengers Injured, 16 and Older, by Vehicle Body Type and Restraint Use, in Single-Vehicle Crashes

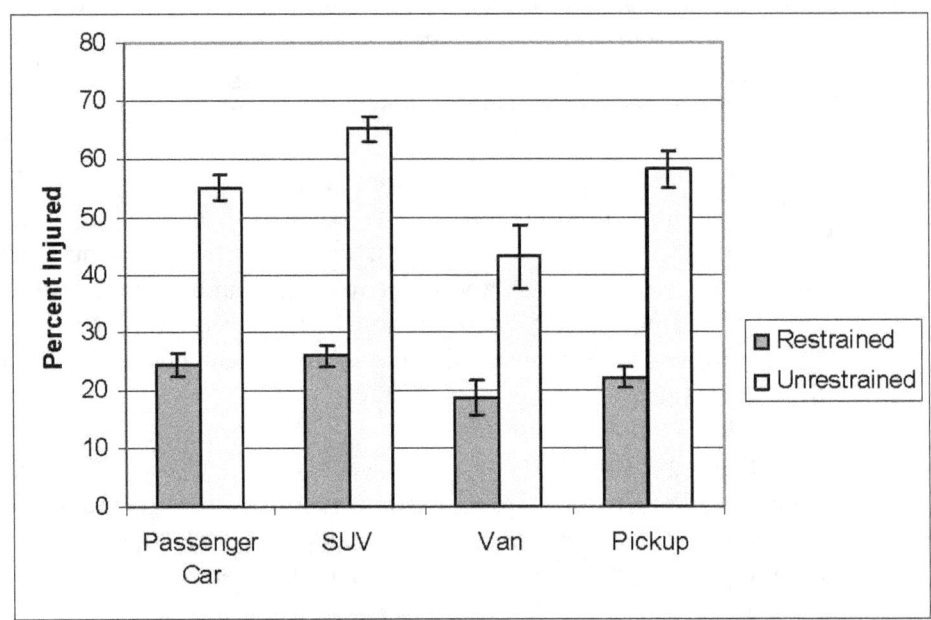

Source: National Center for Statistics and Analysis, NHTSA, GES 1998-2002
Note: Intervals surrounding percent estimates are +/- one standard error. These estimates do not include fatally injured passengers.

7.2 Passengers in Multi-Vehicle Crashes

This section examines passengers who were involved in multi-vehicle non-fatal crashes. Section 7.2.1 focuses on passengers age 0 through 15 and Section 7.2.2 on passengers 16 and older.

Sections 7.1 and 7.2 show that a smaller percentage of unrestrained passengers were injured in multi-vehicle (MV) crashes compared to single-vehicle (SV) crashes. In many types of SV crashes, more than half of the unrestrained passengers were injured, with a high of 65 percent among adult passengers in SUVs; comparatively, the percent of unrestrained passengers in MV crashes who were injured never topped 40 percent.

This comparison suggests that an unrestrained passenger is at a higher risk of being injured when involved in an SV crash versus being involved in a MV crash. This risk difference is due to many factors. For example, compared with MV crashes, vehicles in SV crashes are more likely to experience a rollover or strike a fixed object (i.e. a tree), events that are very likely to injure a passenger in the vehicle. Restrained passengers are also more likely to be injured in an SV crash than in a MV crash, but the difference in risk is much smaller among restrained passengers than among unrestrained passengers.

7.2.1 Passengers Age 0 – 15 in Multi-Vehicle Crashes

In multi-vehicle crashes, 11 to 13 percent of restrained 0 though 15-year-old passengers in LTVs were injured, compared to 17 percent in passenger cars (see Figure 15). For all four vehicle body types, younger passengers in MV crashes were the least likely to be injured of all restrained passengers examined in this chapter, including restrained passengers in single-vehicle crashes (Sections 7.1.1 and 7.1.2), and restrained older passengers in MV crashes (Section 7.2.2).

This is another sign of two factors that are related to a reduced likelihood of injury. First, passengers in multi-vehicle crashes are less likely to be injured compared to passengers in single-vehicle crashes. Second, the younger passengers who are more likely to travel in the safer second seat are less frequently injured compared to those older passengers who are more frequently traveling in the front seat.

As with restrained children, unrestrained children traveling in passenger cars had a higher probability of being injured (37%) than children in SUVs (32%), vans (29%), and pickups (33%). In multi-vehicle crashes, the relative risk of being injured ranged from 2.2 in passenger cars up to 3.0 in SUVs, for unrestrained children compared to restrained children.

Table 13: Percent and Standard Error of Passengers Injured, Age 0 - 15, By Vehicle Body Type and Restraint Use, in Multi-Vehicle Crashes

Body Type	Restrained		Unrestrained	
	Mean (%)	SE (%)	Mean (%)	SE (%)
Passenger Car	16.6	1.1	36.9	2.0
SUV	10.6	1.0	31.8	5.0
Van	12.6	1.0	28.6	3.5
Pickup	12.2	0.9	33.1	2.7

Source: National Center for Statistics and Analysis, NHTSA, GES 1998-2002
Note: These estimates do not include fatally injured passengers.

Figure 15: Percent of Passengers Injured, Age 0 – 15, by Vehicle Body Type And Restraint Use, in Multi-Vehicle Crashes

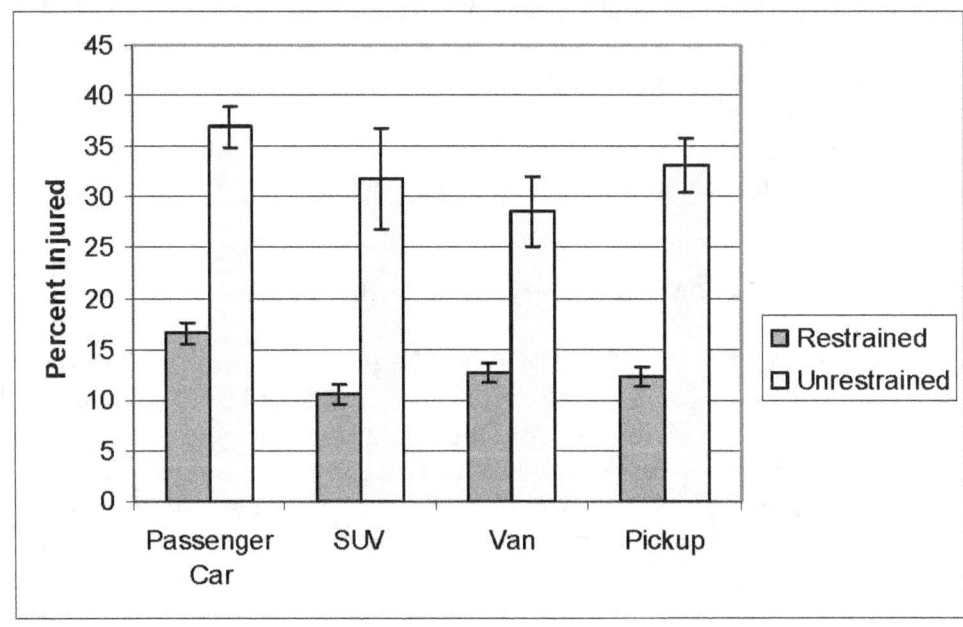

Source: National Center for Statistics and Analysis, NHTSA, GES 1998-2002
Note: Intervals surrounding percent estimates are +/- one standard error. These estimates do not include fatally injured passengers.

7.2.2   Passengers 16 and Older in Multi-Vehicle Crashes

For passengers 16 and older in multi-vehicle crashes, the percent of restrained passengers who were injured was significantly less than the percent of unrestrained passengers who were injured, for all four vehicle body types. This pattern was consistent throughout Chapter 7, showing a clear example of the increased safety provided by restraint use.

As shown in Figure 16, for both restrained and unrestrained passengers, those traveling in passenger cars were the most likely to be injured.

Among restrained passengers, the percent who were injured ranged from 17 percent in pickups up to 23 percent in passenger cars; comparatively for unrestrained passengers, the percent injured was lowest among passengers in vans (31%) and highest in passenger cars (40%).

The relative risk of injury for unrestrained passengers versus restrained passengers was smallest among passengers 16 and older in MV crashes (Section 7.2.2), compared to SV crashes (Section 7.1) or younger passengers in MV crashes (Section 7.2.1). For passengers 16 and older in MV crashes, the relative risk ranged from 1.6 for vans up to 2.1 for SUVs.

Table 14: Percent and Standard Error of Passengers Injured, 16 and Older, By Vehicle Body Type and Restraint Use, in Multi-Vehicle Crashes

Body Type	Restrained		Unrestrained	
	Mean (%)	SE (%)	Mean (%)	SE (%)
Passenger Car	23.3	1.5	39.9	2.1
SUV	17.2	1.6	35.7	3.0
Van	19.0	1.5	31.2	2.5
Pickup	16.5	1.2	31.7	2.5

Source: National Center for Statistics and Analysis, NHTSA, GES 1998-2002
Note: These estimates do not include fatally injured passengers.

Figure 16: Percent of Passengers Injured, 16 and Older, By Vehicle Body Type and Restraint Use, in Multi-Vehicle Crashes

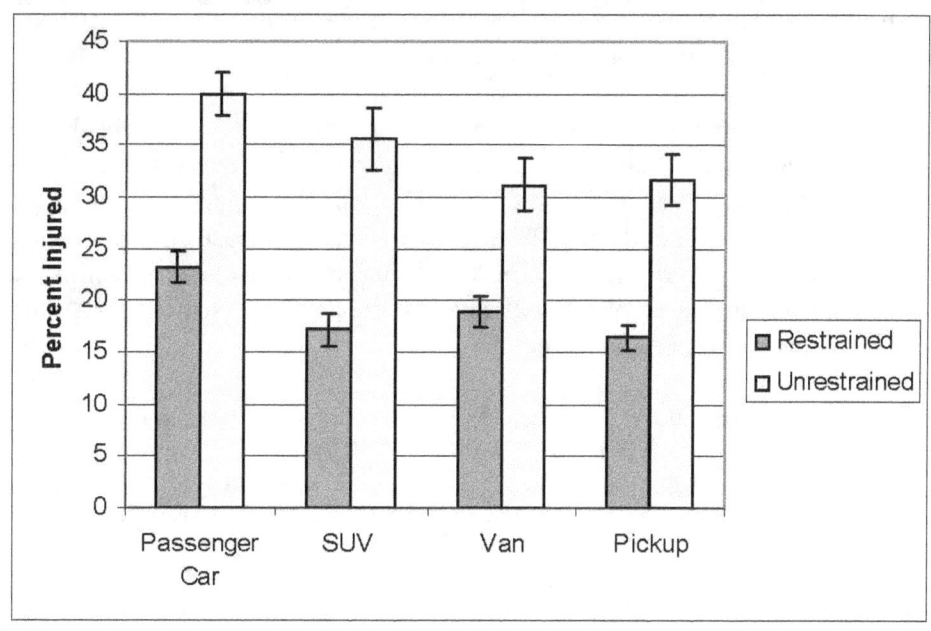

Source: National Center for Statistics and Analysis, NHTSA, GES 1998-2002
Note: Intervals surrounding percent estimates are +/- one standard error. These estimates do not include fatally injured passengers.

8. **Percent of Passengers Fatally Injured, in Two-Vehicle Fatal Crashes Involving Passenger Cars and/or LTVs**

The two-vehicle crashes included in Chapter 8 are categorized by the vehicle body type of the "striking" vehicle and the "struck" vehicle. Vehicle body type for these two-vehicle crashes is separated into two categories: passenger cars (PC) and light trucks and vans (LTV). (LTV is a vehicle group that consists of SUVs, vans, and pickups.) Therefore four crash types are examined:

    i. PC striking PC
    ii. PC striking LTV
    iii. LTV striking PC
    iv. LTV striking LTV

Two-vehicle fatal crashes from 1998 through 2002 are examined. The charts in Chapter 8 show the percent of passengers in the striking vehicle who were fatally injured and the percent of passengers in the struck vehicle who were fatally injured, for each of the four crash types mentioned above. These percentages are also shown for the aggregate of four above-mentioned crash types. Throughout Chapter 8, passengers are separated into four age categories (0-3, 4-7, 8-15, and 16 and older) and two restraint use categories (restrained and unrestrained).

Patterns Seen Among Two-Vehicle Fatal Crashes

Two main patterns were observed within the two-vehicle crash data presented in Chapter 8. These two patterns are seen among both restrained and unrestrained passengers, and through all age categories.

- After aggregating all two-vehicle fatal crashes included in this chapter, a much larger percentage of passengers in the struck vehicle were fatally injured (50%) compared to passengers in the striking vehicle (22%).
- When a PC and an LTV collided in a fatal crash, the passengers in the LTV were less likely to be fatally injured than those traveling in the PC. This finding applies to crashes where an LTV struck a PC, and also to crashes where a PC struck an LTV.

From 1998 through 2002, over 25,000 passengers were fatally injured in two-vehicle crashes involving a PC and/or an LTV. Among these fatalities, 6,609 passengers were killed in the striking vehicle, while 18,457 were killed in the struck vehicle.

In crashes where a PC struck another PC, 21 percent of the passengers in the striking vehicle were killed, while 52 percent of the passengers in the struck vehicle were killed (see Figure 17). For crashes where an LTV struck another LTV, 27 percent of the passengers in the striking vehicle were killed, while 45 percent of the passengers in the struck vehicle were killed. These two types of crashes show that a passenger in the struck vehicle had a smaller chance of surviving a fatal crash than a passenger in the striking vehicle, when the vehicles were of similar body types.

A different pattern is observed when a two-vehicle fatal crash involved a PC and an LTV. Figure 17 shows that when an LTV struck a PC, 60 percent of the passengers in the struck PC were fatally injured, compared to only 9 percent of the passengers in the striking LTV; however, among crashes where a PC struck an LTV, 29 percent of those in the struck LTV were killed, while 45 percent of those in the striking PC were killed.

Figure 17: Percent Killed, Among Passengers in Two-Vehicle Fatal Crashes Involving Passenger Cars and/or LTVs, by Vehicle Role

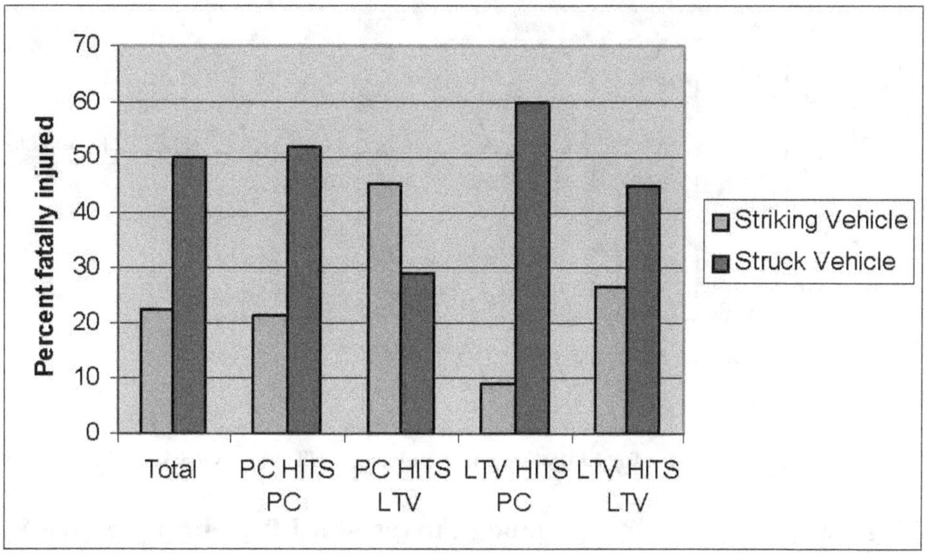

Source: National Center for Statistics and Analysis, NHTSA, FARS 1998-2002

The crash type where a PC strikes an LTV is the only two-vehicle crash type where the passenger was more likely to be fatally injured in the striking vehicle than in the struck vehicle. Potential reasons for this difference include the fact that LTVs frequently possess a larger mass and higher center of gravity than PCs.

Restrained Passengers Compared to Unrestrained Passengers

While Figure 17 presents findings on all passengers in two-vehicle crashes, parallel information on restrained passengers and unrestrained passengers is shown below in Figures 18 and 19, respectively.

For restrained passengers in two-vehicle fatal crashes, Figure 18 shows that 46 percent of those in the struck vehicle were fatally injured, compared to 14 percent in the striking vehicle. Among unrestrained passengers, Figure 19 shows these percentages to be much higher, with 37 percent of those in the struck vehicle fatally injured, compared to 58 percent in the striking vehicle.

These numbers show that whether the passenger is unrestrained or restrained, the chance of being killed in a fatal crash is much higher among passengers in the struck vehicle than passengers in the striking vehicle; however this difference is smaller among unrestrained passengers (21 percentage points difference: 58% versus 37%) than among restrained passengers (32 percentage points difference: 46% versus 14%).

Figure 18: Percent Killed, Among Restrained Passengers in Two-Vehicle Fatal Crashes Involving Passenger Cars and/or LTVs, by Vehicle Role

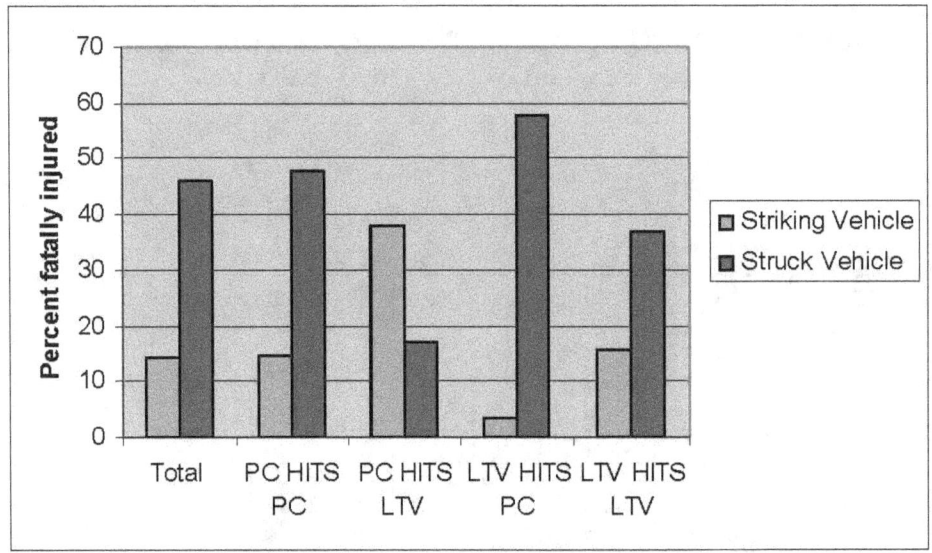

Source: National Center for Statistics and Analysis, NHTSA, FARS 1998-2002

Figure 19: Percent Killed, Among Unrestrained Passengers in Two-Vehicle Fatal Crashes Involving Passenger Cars and/or LTVs, by Vehicle Role

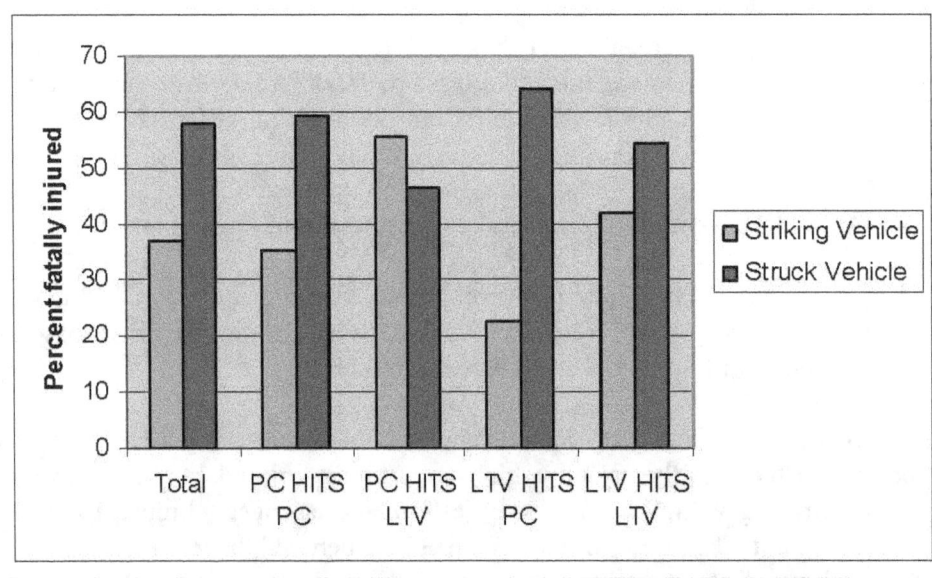

Source: National Center for Statistics and Analysis, NHTSA, FARS 1998-2002

8.1   Passengers in Two-Vehicle Fatal Crashes, Age 0 - 3

This section examines child passengers up through 3 years old who were involved in two-vehicle fatal crashes. Section 8.1.1 focuses on restrained passengers and Section 8.1.2 on unrestrained passengers.

8.1.1   Restrained Passengers, Age 0 - 3

Among restrained children up through 3 years old in two-vehicle fatal crashes, 12 percent of the passengers in the striking vehicle were killed, while 28 percent of the passengers in the struck vehicle were killed.

Figure 20 shows that when an LTV struck a PC, 35 percent of the passengers in the struck PC were fatally injured, compared to only 2 percent of the passengers in the striking LTV; however, among crashes where a PC struck an LTV, 13 percent of those in the struck LTV were killed, while 33 percent of those in the striking PC were killed. Among these two types of fatal crashes that involved a PC and LTV, restrained children were less likely to be fatally injured in an LTV versus a PC.

Figure 20: Percent Killed, Among Restrained Passengers Age 0-3, in Two-Vehicle Fatal Crashes Involving Passenger Cars and/or LTVs, by Vehicle Role

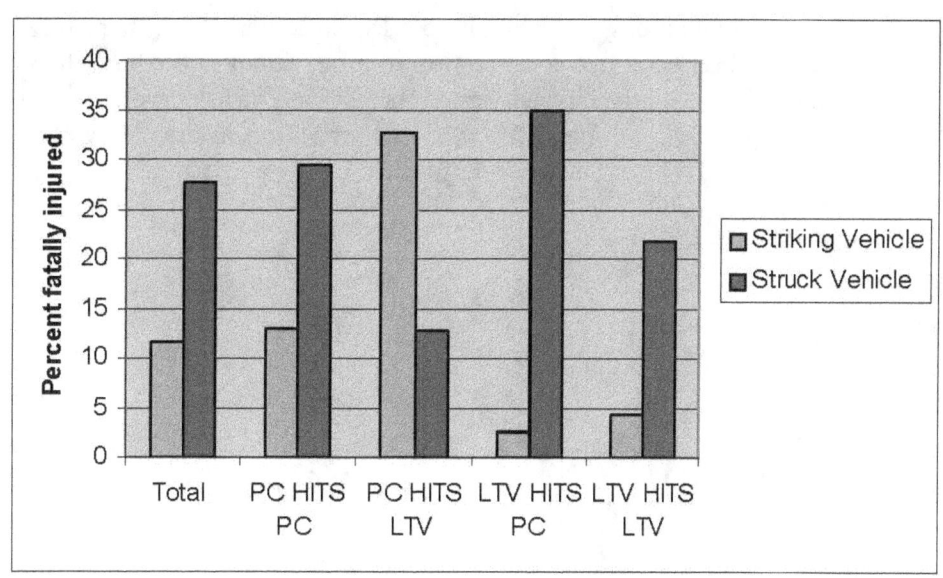

Source: National Center for Statistics and Analysis, NHTSA, FARS 1998-2002

8.1.2 Unrestrained Passengers, Age 0 - 3

While Figure 20 displays data on restrained passengers, Figure 21 (shown below) displays parallel information on unrestrained passengers. A comparison of these two charts shows that when the child passengers were unrestrained in a two-vehicle fatal crash, they were much more likely to be fatally injured than when they were restrained. For unrestrained children up through 3, 40 percent of the child passengers in the striking vehicle were killed, while 46 percent of the passengers in the struck vehicle were killed.

These fatality percentages were larger than those provided in Section 8.1.1, which represent the percent of restrained children who were fatally injured in the striking vehicle (12%) and struck vehicle (28%). The smaller percent of restrained young passengers who were killed, as shown in Figure 20, is an example of the large safety benefit of being properly restrained.

When the child was unrestrained, vehicle body type played less of a role in determining a child's chance of surviving a two-vehicle crash, than when the child was restrained. This relationship parallels what has been shown throughout this report, which is that being unrestrained removes much of the safety protection that vehicles of certain body types can provide a restrained child passenger.

Other than among crashes where a PC struck an LTV, there was only one age and restraint use combination in all of Chapter 8 where the percent of passengers fatally injured was higher among striking vehicles (56%) than among struck vehicles (37%). This was seen in crashes where an LTV struck an LTV, as shown in Figure 21.

Figure 21: Percent Killed Among Unrestrained Passengers Age 0-3, in Two-Vehicle Fatal Crashes Involving Passenger Cars and/or LTVs, by Vehicle Role

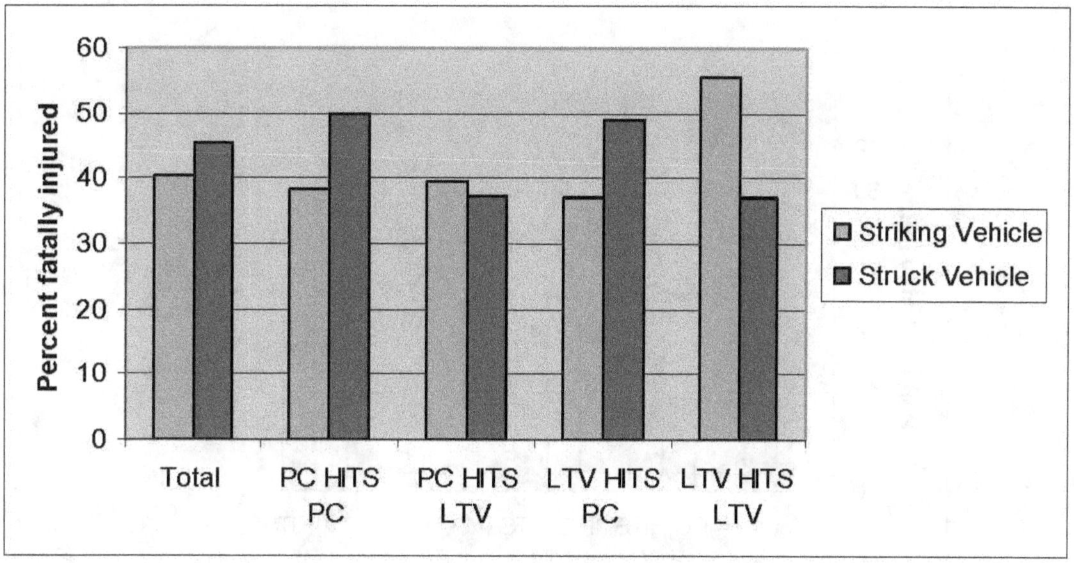

Source: National Center for Statistics and Analysis, NHTSA, FARS 1998-2002

8.2   Passengers in Two-Vehicle Fatal Crashes, Age 4 - 7

This section examines child passengers age 4 through 7, who were involved in a two-vehicle fatal crash.   Section 8.2.1 focuses on restrained passengers and Section 8.2.2 on unrestrained passengers.

8.2.1   Restrained Passengers, Age 4 - 7

Among restrained children 4 through 7 years old in two-vehicle fatal crashes, 7 percent of the passengers in the striking vehicle were killed, while 24 percent of the passengers in the struck vehicle were killed. Percentages similar to these were seen among crashes where a PC struck a PC, and where an LTV struck an LTV, as shown in Figure 22.

In crashes where an LTV struck a PC, the percent of restrained children in the LTV who were killed (3%) was less than one-tenth as high as the percent of restrained children in the PC who were killed (33%); yet even when a PC struck an LTV, the percentage of restrained children in the PC who were killed (19%) was more than twice as high as the percent of restrained children in the LTV who were killed (9%).

Figure 22: Percent Killed, Among Restrained Passengers Age 4-7, in Two-Vehicle Fatal Crashes Involving Passenger Cars and/or LTVs, by Vehicle Role

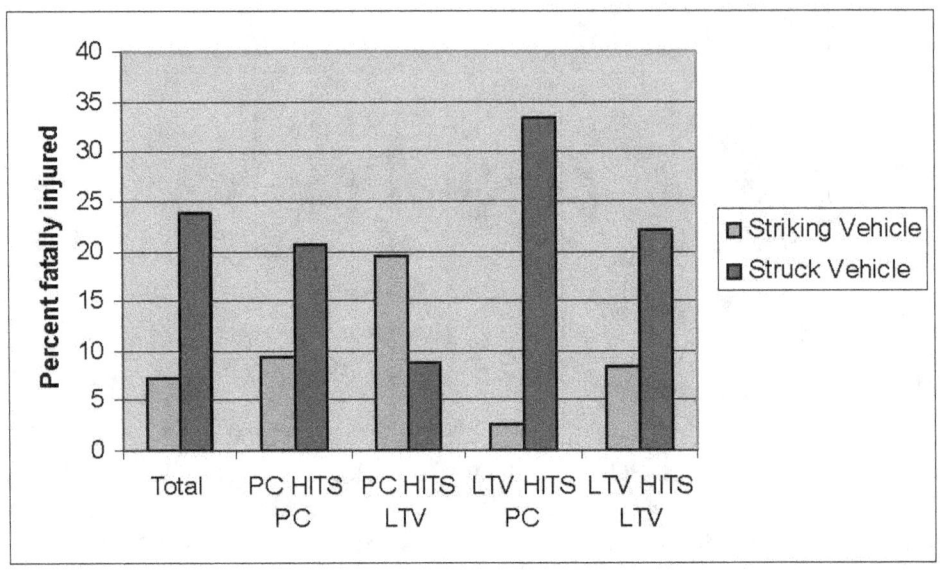

Source: National Center for Statistics and Analysis, NHTSA, FARS 1998-2002

8.2.2 Unrestrained Passengers, Age 4 - 7

For unrestrained children age 4 through 7, 24 percent of the child passengers in the striking vehicle were killed, while 33 percent of the passengers in the struck vehicle were killed. Figure 23 shows that when the child passengers were unrestrained in a two-vehicle fatal crash, they were much more likely to be fatally injured than when they were restrained.

This difference in the percentage of unrestrained children fatally injured in the striking vehicle (24%) versus the struck vehicle (33%) was only half as large the difference seen among restrained children (7% versus 24%). This is another example that suggests that when the child is unrestrained, distinguishing factors in the type of crash play less of a role in whether a child survives the crash.

Figure 23: Percent Killed, Among Unrestrained Passengers Age 4-7, in Two-Vehicle Fatal Crashes Involving Passenger Cars and/or LTVs, by Vehicle Role

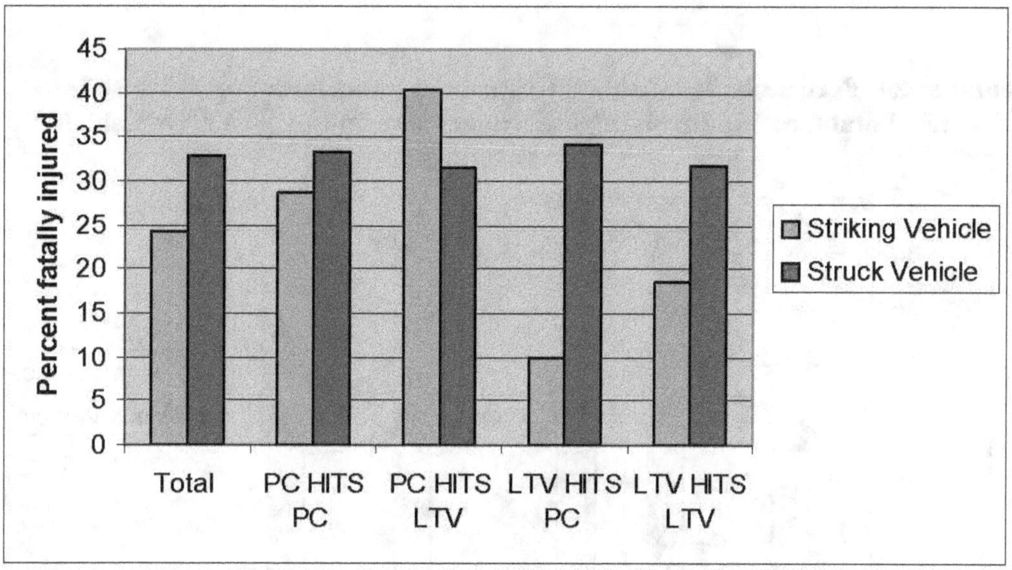

Source: National Center for Statistics and Analysis, NHTSA, FARS 1998-2002

8.3   Passengers in Two-Vehicle Fatal Crashes, Age 8-15

This section examines child passengers age 8 through 15 who were involved in two-vehicle fatal crashes. Section 8.3.1 focuses on restrained passengers and Section 8.3.2 on unrestrained passengers.

8.3.1   Restrained Passengers, Age 8 - 15

Figure 24 shows that when an LTV struck a PC, 32 percent of the passengers in the struck PC were fatally injured, compared to only 1 percent of the passengers in the striking LTV; however, in crashes where a PC struck an LTV, only 6 percent of those in the struck LTV were killed, while 16 percent of those in the striking PC were killed. Among these two types of fatal crashes that involved a PC and LTV, restrained children are less likely to be fatally injured in an LTV versus a PC.

In crashes where a PC struck a PC, the children in the struck PC were 7 times as likely to be fatally injured as the children in the striking PC; in LTV-LTV crashes the children in the struck vehicle were just over 2 times as likely to be fatally injured in the struck LTV compared to the striking LTV.

Among all restrained children 8 through 15 in two-vehicle fatal crashes, 5 percent of the passengers in the striking vehicle were killed, while 23 percent of the passengers in the struck vehicle were killed.

Figure 24: Percent Killed, Among Restrained Passengers Age 8-15, in Two-Vehicle Fatal Crashes Involving Passenger Cars and/or LTVs, by Vehicle Role

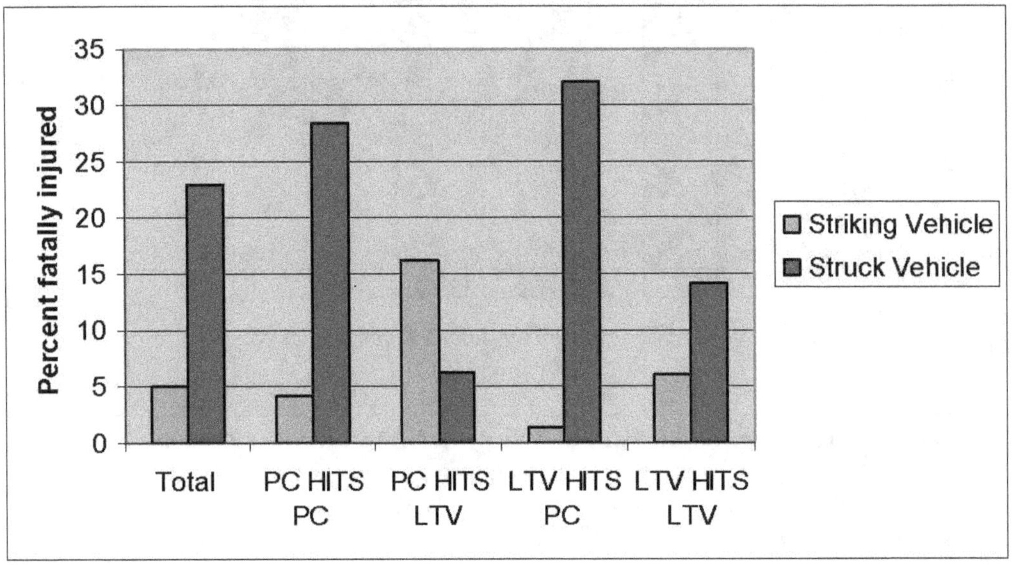

Source: National Center for Statistics and Analysis, NHTSA, FARS 1998-2002

8.3.2   Unrestrained Passengers, Age 8-15

When a PC struck an LTV, 30 percent of the unrestrained child passengers in the PC were fatally injured, as were 24 percent of the unrestrained passengers in the LTV. In the each of the other three crash categories in Figure 25 (PC strikes PC, LTV strikes PC, and LTV strikes LTV), unrestrained children in the striking vehicle were less likely to be killed (ranging from 12 to 18%) than were unrestrained children in the struck vehicle (ranging from 30 to 43%).

Among all unrestrained children 8 through 15 years old in two-vehicle fatal crashes, 18 percent of the passengers in the striking vehicle were killed, while 33 percent of the passengers in the struck vehicle were killed. These percentages were much higher than the comparable percentages seen among restrained children age 8 through 15 (5 and 23%).

Figure 25: Percent Killed, Among Unrestrained Passengers Age 8-15, in Two-Vehicle Fatal Crashes Involving Passenger Cars and/or LTVs, by Vehicle Role

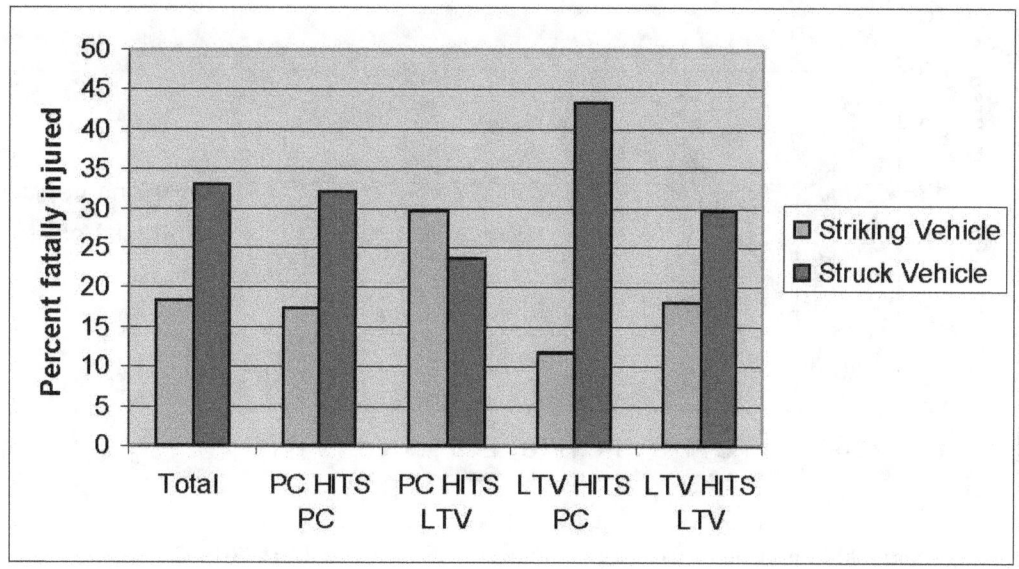

Source: National Center for Statistics and Analysis, NHTSA, FARS 1998-2002

8.4   Passengers in Two-Vehicle Fatal Crashes, 16 and Older

This section examines passengers 16 and older who were involved in a two-vehicle fatal crash. Section 8.4.1 focuses on restrained passengers and Section 8.4.2 on unrestrained passengers.

8.4.1   Restrained Passengers, 16 and Older

In fatal crashes where an LTV struck a PC, only 225 restrained passengers in the LTV were killed and 5,992 survived, while 5,342 passengers were killed in the PC, with 3,377 surviving. These numbers show that a far higher percentage of adult occupants in the struck PC were fatally injured (61%) compared to the striking LTV (4%).

When a PC struck an LTV, restrained passengers in the PC were more than twice as likely to be fatally injured (40%) compared to restrained passengers in the LTV (19%). As these percentages show in Figure 26, passengers in an LTV are more likely to survive a fatal crash with a PC, regardless of which vehicle is the striking vehicle.

Figure 26: Percent Killed, Among Restrained Passengers Age 16 and Older, in Two-Vehicle Fatal Crashes Involving Passenger Cars and/or LTVs, by Vehicle Role

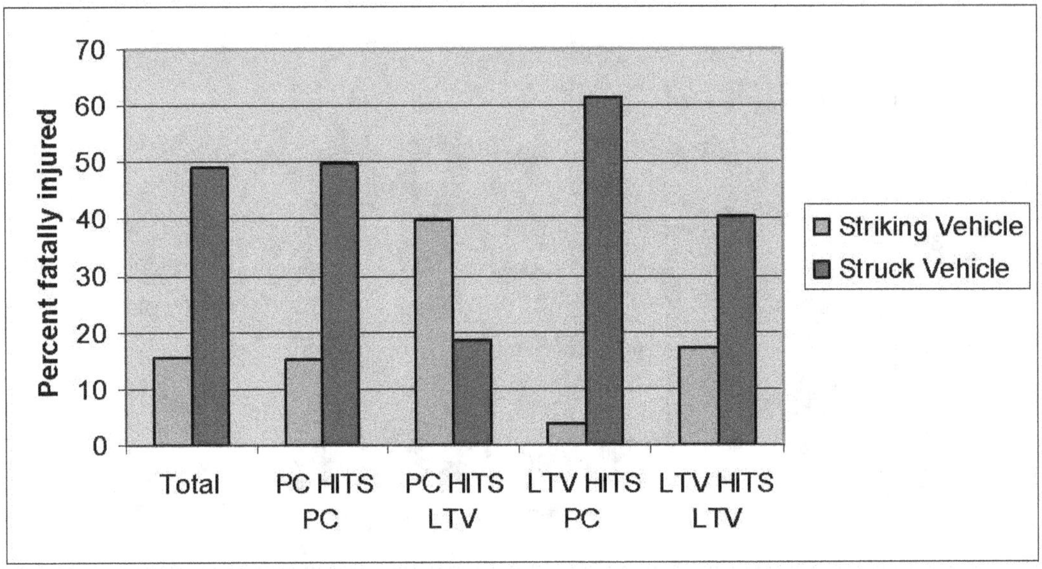

Source: National Center for Statistics and Analysis, NHTSA, FARS 1998-2002

8.4.2   Unrestrained Passengers, 16 and Older

Figure 27 suggests that when occupants 16 and older were unrestrained in two-vehicle fatal crashes, they were more likely to be fatally injured than if they were restrained. Among unrestrained occupants, 39 percent of the occupants in the striking vehicle were killed, while 61 percent of the occupants in the struck vehicle were killed.

These fatality percentages were larger than those shown in Section 8.4.1 that represent the percent of restrained occupants who were fatally injured in the striking vehicle (15%) and struck vehicle (49%).

Among unrestrained adult passengers, vehicle type played less of a role in determining a passenger's chance of surviving a two-vehicle crash, compared to when a passenger was restrained. For example, in crashes where an LTV struck a PC, 4 percent of restrained passengers in the LTV were killed and 61 percent of restrained passengers in the PC were killed. These percentages differed less among unrestrained passengers, with 23 percent of those in the striking LTV being killed, compared to 67 percent in the struck PC.

Figure 27: Percent Killed, Among Unrestrained Passengers 16 and Older, in Two-Vehicle Fatal Crashes Involving Passenger Cars and/or LTVs, by Vehicle Role

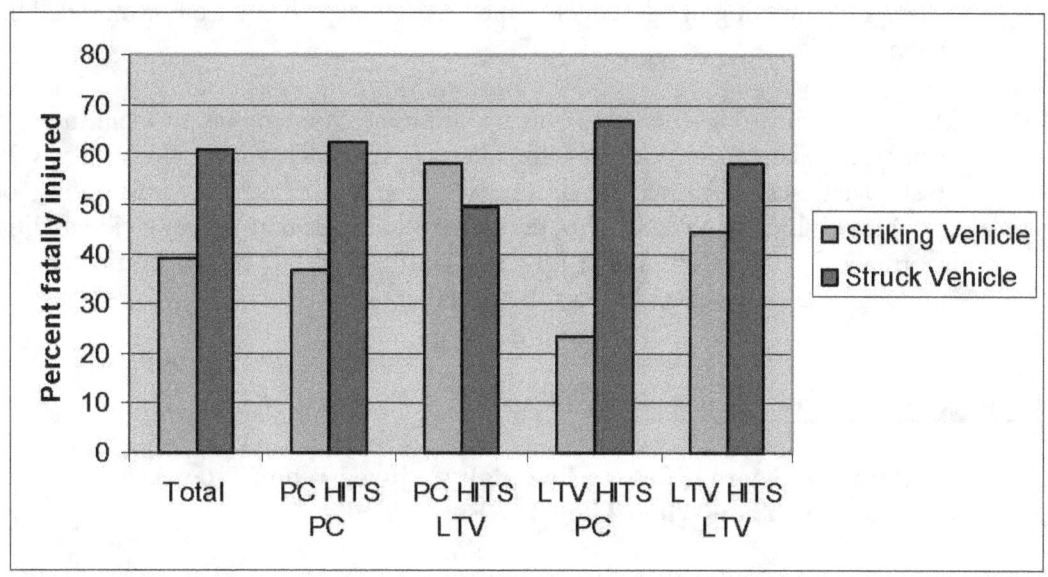

Source: National Center for Statistics and Analysis, NHTSA, FARS 1998-2002

9.   Conclusions

The findings of this report support a variety of conclusions about the safety of passengers traveling in passenger vehicles. Vehicle drivers are not included in the analysis.

This report shows the benefit of using different data sources to examine similar variables. The report does not consider all variables within the motor vehicle crash databases. Further studies need to be undertaken by examining other variables within the FARS, GES, and other databases that may provide additional information. The National Highway Traffic Safety Administration plans to conduct these analyses and report the findings.

Analysis of Fatal Crashes (9.1 – 9.6)

Children Are More Likely to Be Fatally Injured When They Are Unrestrained Than When They Are Restrained

In single-vehicle fatal crashes, the relative risk of an unrestrained child in a passenger car being fatally injured was between 2.0 and 2.3, compared to restrained children. In LTVs, this relative risk ranged from 1.9 for 0-3 year olds in vans, up to 3.1 for 8-15 year olds in pickups.

In multi-vehicle fatal crashes, unrestrained children in LTVs were 2.5 to 5.4 times as likely to be fatally injured as children who were restrained. The relative risk was highest among children in SUVs and lowest for children in vans. Comparatively, the relative risk was smaller for children in passenger cars than for children in LTVs. For a child in a passenger car in a multi-vehicle crash, being unrestrained made the child between 1.6 and 1.8 times as likely to be fatally injured, compared to a restrained child.

In Fatal Crashes, the Percent of Restrained Child Passengers Who Are Fatally Injured Is Higher in a Passenger Car Than in an LTV

In single-vehicle fatal crashes, 20 percent of restrained children age 0-3 in passenger cars were fatally injured. A smaller percentage of children in SUVs (11%), vans (14%), and pickups (15%) were fatally injured. This pattern among single-vehicle fatal crashes was also seen among the older age groups. Restrained children overall were 1.5 times as likely to be fatally injured in a passenger car compared to an LTV.

In multi-vehicle fatal crashes, restrained children in passenger cars were roughly 2.5 times as likely to be fatally injured as restrained children in LTVs, a relative risk higher than what was seen among single-vehicle fatal crashes. These statistics raise the issue of vehicle compatibility.

When a Child Is Unrestrained, Vehicle Body Type Is Less of a Determinant of a Child's Chance of Survival, Than When a Child Is Restrained

Although unrestrained children in passenger cars were more likely to be fatally injured than unrestrained children in LTVs, this difference is less than the difference seen among restrained children (as discussed in Conclusion 9.2). The percentage of unrestrained children who were killed was more homogeneous, among different vehicle body types, than the percentage of restrained children killed. Being unrestrained thus removes much of the safety protection that a vehicle can provide a restrained child passenger.

Children Are Safer When Traveling in the Second Seat

Whether children are restrained or unrestrained, they are safer when traveling in the second seat. In fatal crashes, restrained children in the front passenger seat were roughly 1.5 times as likely to be fatally injured compared to restrained children in the second seat. The relative protection provided by traveling in the second seat (compared to the front passenger seat) is lessened when the passenger is unrestrained.

Among Most Two-Vehicle Fatal Crashes, the Passengers in the Struck Vehicle Are More Likely to Be Killed Than Those in the Striking Vehicle

From 1998 through 2002, over 25,000 passengers of all ages were killed in two-vehicle crashes involving passenger cars and/or LTVs. Among these fatalities, nearly 3 times as many occurred in the struck vehicle (18,457 passengers) compared to the striking vehicle (6,609 passengers).

Whether the passenger is unrestrained or restrained, the chance of being killed in a fatal crash is higher among passengers in the struck vehicle than passengers in the striking vehicle; however, this difference is smaller among unrestrained passengers than among restrained passengers.

In Two-Vehicle Fatal Crashes Between a Passenger Car and an LTV, the Vehicle Body Type Plays a Role in the Passenger's Chance of Survival

When limiting the scope of two-vehicle fatal crashes to those between a passenger car and an LTV, a strong pattern emerged. In these crashes, passengers in the LTV were far more likely to survive the fatal crash than passengers in the passenger car, whether the LTV struck the passenger car or the passenger car struck the LTV.

<u>**Analysis of Non-Fatal Crashes (9.7 – 9.8)**</u>

Unrestrained Passengers Are Much More Likely to Be Injured Than Restrained Passengers

An examination of non-fatal crashes showed that unrestrained passengers were more likely to be injured than were restrained passengers. This difference was statistically significant regardless of the age of the passenger, the body type of the vehicle in which the passenger was traveling, and/or the number of vehicles in the crash.

Unrestrained Passengers Are at a Higher Risk of Being Injured When in a Single-Vehicle Crash Versus in a Multi-Vehicle Crash

The probability of unrestrained passengers being injured was consistently larger among single-vehicle crashes than among multi-vehicle crashes. In many types of single-vehicle crashes, a majority of unrestrained passengers were injured.

Among unrestrained passengers in single-vehicle crashes, those age 16 and older traveling in SUVs were the most likely to be injured (64%), while those up through 15 years old were the least likely to be injured (43%). In multi-vehicle crashes, by comparison, the percent of unrestrained passengers who were injured never topped 40 percent.

www.ingramcontent.com/pod-product-compliance
Lightning Source LLC
Chambersburg PA
CBHW081615170526

45166CB00009B/2977